Date Due

Here is truth that beats fiction!
Old Panamint's silver hoard was
first found by stage-robbers who
had fled to the heights above
Death Valley to escape sheriffs and
marshals. It was rediscovered by
prospectors searching for a lost
mine. On the heels of the rumor
of a new bonanza the miner, the
merchant, the courtesan, the gambler, the gunman all rushed to Panamint. Uncle "Billy Bedamned"
Wolsesburger, aged peddler, limped
417 miles over the deserts whacking his little burro. John Schober
crossed 166 miles with a big
whipsaw on his back. Clem Ogg,
who could cut the seat out of a
man's trousers with the lash of his
long bull whip, hitched fourteen
freight wagons behind a half-mile
parade of bullocks and set out. Into

(*Continued on back flap*)

THE MACMILLAN COMPANY

*Publishers*                    *New York*

(*Continued from front flap*)

the gulch town that sprang up strode George Hearst, Lucky Baldwin, Senators J. P. Jones and "Fifteenth Amendment" Bill Stewart, and many other famous characters of the rough old West who later brushed elbows at Bodie, Tombstone and the Black Hills.

Wells Fargo, bullion carrier for every mining settlement between the Sierra Nevada and the Missouri River, refused to have anything to do with this "doorstep to hell." How Senator Bill Stewart got his treasure down the narrow corridor and out across the deserts with the highwaymen dumfoundedly watching it go, is one of the humorous climaxes of this chronicle. How Stage-Driver Jack Lloyd found life one vast joke after another, and how it finally treated him; how Fred Yager imported the biggest bar mirror in the Seven Deserts, and what came of it; the imposing entrance of Martha Camp and her impious damsels—these and the adventures of innumerable sagebrushers, diehards and rawhided old-timers of the Silver Seventies are told by Mr. Wilson with full appreciation of the picturesque, the zestful, the humorous, and the violent.

# SILVER STAMPEDE

## THE CAREER OF DEATH VALLEY'S HELL-CAMP, OLD PANAMINT

BY

NEILL C. WILSON

*ILLUSTRATED*

NEW YORK
THE MACMILLAN COMPANY
1937

PRINTED IN THE UNITED STATES OF AMERICA
NORWOOD PRESS LINOTYPE, INC.
NORWOOD, MASS., U.S.A.

To
WALTER J. WILSON

# MARCH OF THE OLD–TIMERS
## [DEATH VALLEY]

*Pale the salt sink winds beneath the stars,*
*A silent street between dark doorless walls.*
*Wary we'll peer from hill-hung balcony*
*Into its glimmer.*

                 *See, with healing night*
*Figures once bold long broken come astir.*
*Here beside dust hole, bright-ribbed skeleton*
*Belabors spectral jack that brays the cliffs*
*With soundless laughter. Yonder, skirting lake*
*That dances in illusion, mother-wraith*
*In lank-flung hair and crumbling calico*
*Calls to her wandered children.*

                 *Windblown mounds*
*Crack and upheave. Gaunt cakebeards start and stretch:*
*Sagebrushers, sourdoughs, diehard desert rats,*
*Rawhide old-timers unallowed to rot,*
*The muster of the nameless. Brittle shapes*
*With swollen tongues, and rheumy eyes turned in,*
*Drag they to foot, and finding pavement cooled,*
*Cross and recross that way of dreadful glint.*
*With shriveled fists they pound those mocking walls,*
*Venting stale blasphemies of rage, doubt, fright,*
*Torment, horror. No latch yields.*

                 *When now*
*Rises the moon. Our gravebirds wheel about.*
*Grinning last-chancers, losers on the draw,*
*The column of the beaten, phantom files*

Shouldering packs and shovels white with crust,
Brandishing whisky bottles long since drained,
Form they sear ranks, and under silver orb
March for the saline well.  Jocose they dip
Into its bitter wassail, cynic skulls
Cackling each straggler, hooting each hoar tale
Of lode near-found, bonanza almost staked,
Strike within sound of faint but certain stream,
Ere the consuming noonblast.  Still they come:
The corpse of him who tried the shorter route;
The stumbling shade whose mule-span hightailed off;
The trapped who fleeing would not drop his gold;
And he who brained his pardner, arm in arm
With him who took the stoving.  Men destroyed,
Toast they each other and the victor sun.

When from our alpine lookoff we behold
Red-edged that sun.  Our dead men scurry swift,
Each to his shallow sepulcher, to wait
Renewed assembly when the sand grows kind.

# CONTENTS

# ACKNOWLEDGMENTS

FOR this reconstruction of the career of a ghost town of the silverlands, with strict attention to the basic outlines and liberal interpolations to preserve the flavor of those racy times, I am under deep obligation to Mrs. Georgina Sullivan Jones, who looked back over sixty intervening years; to Mr. Roy Jones, the Senator's son, for private photographs; to W. A. Chalfant, the Sage of Inyo, who provided me with his three cherished copies of the Panamint *News* and other information; to that monumental work of his father, the late Pleasant A. Chalfant, pioneer, whose *Inyo Independent* in its brittle-brown files still hints the industry and tolerance that motivated this conscientious frontier journalist; to William A. Irwin, a grand old-timer of the Panamint hills, who contributed personal knowledge to these sketches; to J. Vincent Lehigh of Los Angeles, title-holder today to much of Surprise Valley's latent silver; to sturdy Chris Wicht who dwells in his cabin at the foot of Surprise Canyon, and has often showed me where to fling my blankets; to James L. Bateman, Chris' occasional neighbor; to Mrs. Charlotte Hibberd, who as little Charlotte Vickers once romped at Panamint while the guns barked outside the *News* office and her father, Rome G. Vickers, set up the type for the indicated obituary; to Mrs. George B. Jacobs and Richard Jacobs, daughter and grandson respectively of the co-founder of Panamint; to Philip Johnston for data and relics; to Gabriel Moulin, Raymond Moulin, Albert Geisecke and E. A. Burbank for illustrations; to the Bancroft, California State, California Historical Society and Society of California Pioneers libraries; to Theodore J. Hoover, Dr. Aurelio M. Espinosa and Mrs. Maurine Marsh for counsel; to Miss Dorothy Huggins, Mrs. Rogers Parratt, Miss Vivian Yarbrough, Miss Leola Fredrickson and Mrs. Elsie M. Rosa for researches; to Bestor Robinson, Horace H. Breed, Harrison Ryker, Lee S. Stopple, Carl I. Wheat, Leon O. Whitsell, Edgar M. Kahn, Oliver Kehrlein, Frank Olmstead, Edgar H. Bennett, G. H. Stern, R. L. Skinner, Charles Bowman, Norman Clyde, Phil Townsend Hanna, and all other fellow-explorers; to Gladys P. Wilson and Walter J. Wilson for happy campfire companionship; and to all those Sourdoughs, Sagebrushers and Rawhide Old Timers who have halted their jacks and stayed their fry-pans to ransack their recollections and helped me to reproduce this desert sideshow.

Other material for this record was found chiefly in the following:

Bakersfield *Courier, Southern Californian*; Bodie *Free Press, Standard*; Darwin *Coso Mining News*; Eureka *Sentinel*; Independence *Inyo Weekly Independent*; Los Angeles *Express, Star, Herald*; Panamint *News*; Pioche *Record*; Sacramento *Bee, Union, Record-Union*; San Bernardino *Argus, Guardian*; San Francisco *Chronicle, Bulletin, Examiner, Post, Call, Daily Alta California*; Santa Monica *Outlook*; Virginia City *Enterprise, Chronicle*.

Daggett, John: *Scrapbooks*; California State Library, Sacramento. Hume, James B.: *Scrapbooks and personal files*; Wells Fargo Bank & Union Trust Co., San Francisco. Smith, Herbert Lee, *The Bodie Era*; MS. in Bancroft Library, University of California, Berkeley, and California State Library, Sacramento.

Banning, Captain William and Banning, George Hugh: *Six Horses*; New York, 1930. Bell, Horace: *Reminiscences of a Ranger; or, Early Times in Southern California*; Los Angeles, 1881. Bell, Horace: *On the Old West Coast*, Edited by Lanier Bartlett; New York, 1930. Burdette, Robert J.: *Greater Los Angeles and Southern California*; Los Angeles. Carr, John: *Pioneer Days in California*; Eureka, 1891. Chalfant, W. A.: *Death Valley, the Facts*; Stanford University, 1933. Chalfant, W. A.: *Outposts of Civilization*; Boston, 1928. Chalfant, W. A.: *The Story of Inyo*; Bishop, California, 1933. Davis, Sam. P., Ed.: *History of Nevada*, 2 vols.; Reno and Los Angeles, 1913. Dobie, J. Frank: *Coronado's Children*: Garden City and New York, 1930. Glasscock, Carl Burgess: *The Big Bonanza: The Story of the Comstock Lode*; Indianapolis, 1931. Goodwin, C. C.: *As I Remember Them*; Salt Lake City, 1913. Jaeger, Edmund C.: *The California Deserts*; Stanford University, 1933. Lord, Eliot: *Comstock Mining and Miners*; Washington, 1883. Lyman, George D.: *The Saga of the Comstock Lode*; New York and London, 1934. Manly, William Lewis: *Death Valley in '49*; New York and Santa Barbara, 1929. Newmark, Harris: *Sixty Years in Southern California*; Boston and New York, 1930. *Pacific Coast Annual Mining Review and Stock Ledger*; San Francisco, 1878. Stewart, William M.: *Reminiscences*, edited by George Rothwell Brown; New York and Washington, 1908.

# ILLUSTRATIONS

xiii

# SILVER STAMPEDE

1

## SIXTY YEARS AFTER

WE press against the sagging door. Its hinge resists. But
we are resolute intruders. We force the red complaining
strap and shove back the inward rubbish. A slumped hearth,
spilling ashes. Whose sour dough once found fullness in its
warmth? An old boot, split and brittle. How far did it come
trudging, across what sands, after what mirages? A slimmer,
lighter, altogether sprightlier foot-covering. Did its missing
mate run away with the mate of the brogue, leaving this duo
to their dissolute housekeeping? A miner's pick. This too
has traveled, for it is foundry-stamped "Washoe" beneath
the thick brown rust.

And three or four cases of broken champagne bottles.

Well, life was here while it lasted. But all was long ago.
The roof is gone. Rafters of knotty juniper hang in inverted
V's, through which stabs the sun, hard and swift as Bill
McLaughlin's knife. That heap over in the corner must
have been the bed. Martha Camp's, conceivably, toward
which Mr. Barstow once stormed with gun in hand, while
Mr. Bruce rolled out on the other side and coolly waited—
the frightened Sophie Glennon, in her unchaste shift, standing here by the doorway meanwhile with the lamp. Or was
the veteran couch Mr. Bruce's own, and this his cabin, whither
Mr. Barstow was fetched after the battle to be decently laid

I

out—Mr. Bruce being not only a dealer of both faro and fate but a neat and handy undertaker as well?

Yonder, through the unstopped window, what is left of the street sprawls steeply up on the east toward a rose-pink granite wall, and on the west steeply down and over a brink. Beyond the wall and below the brink are sternest deserts.

A queer place for a town, this elevated gorge. It is more like a trap. Yet we begin to make out the square stone patterns of forgotten houses. There are fully two hundred of them. They jostle in a continuous row up and down both sides of the gulch; in its day each cabin evidently leaned chummily against its neighbor. Now they lie sprawled like drunks that trusted their own legs not at all and each other's too well. Occasionally a stump of chimney leers through the rabbit brush.

Which of these battered foundations once supported Dave Neagle's Oriental? Which was Fred Yager's Dexter, with Bob Peterson behind the bar and ready to duck? Which was Joe Harris' resort where a Chinaman couldn't wash windows without risking a shot between the legs? Was this the stage office in front of which Jim Bruce and Bob McKenny had it out, while the secretary and a distinguished member of the United States Senate dropped off the coach in a hurry and down behind its wheels? Which of those zig-zag trails on the mountain did Small and McDonald take after making their business calls on the bank?

Here in this wide vacant street once swung the fourteen- and twenty-four-mule teams of Newt Noble and Remi Nadeau. Up from down yonder tooled Bill Buckley's stage-coaches, bullet-punctured now and then and dust-caked always. Down this spiraling corridor and out onto the salt and sagebrush moved some notable freight-wagons, with

*Drawn by E. A. Burbank.*

Panamint today, looking down-canyon.

*Gabriel Moulin photo.*

Panamint today, looking up Surprise Valley.

MAIDEN LANE, 1936

Neagle's Oriental in left foreground; Martha Camp's was somewhere beyond.

"Here men lived and hustled and hoped."

most of their mules hitched on behind for pullback purposes. From the bright snows over our heads to the simmering trough down there lies a two-mile plunge through violent scenery. The cargoes that those mules held back sometimes consisted of quarter-ton chunks of silver.

Cloudbursts of six decades have roared through the gorge that cupped this town. Yet its ground-design clings. Here men lived and hustled and hoped, and after all was over, here some sat on in their doorways and just waited. One can do a lot of waiting in this country, though split-seconds also have had their importance. Up in Sour Dough Canyon, a side ravine, fifty-seven men lie sepulchered, a good many in their boots, and one wonders who dumped the last one in.

Though it was never more than a sideshow to the drama of the Silver Seventies, the career of this camp must have contained much that was typical and a little that was unique. An old newspaper is pasted here on the wall. Brown and fragile, no longer keeping out any wind for the ghostly occupants, it is an Anaheim *Gazette* of November, 1874. States a paragraph ringed by a forgotten hand:

"There are 700 men, 10 women and 4 inches of snow up at Panamint, and lively times are expected."

# THE LIGHT ON THE DESERT

On a rocky desert outthrust a man and a boy crouched and watched. They were not easy as they watched. But they were silently absorbed. The aboriginal tongue had no words for what had pushed into sight below. Down through a gash in the opposite mountains the strange procession was moving: horned creatures bigger than mountain sheep, and yoked two-by-two; house-like structures that followed close, tilting and lurching; men with hairy faces who waved their arms and wielded whips as they walked.

It was December, 1849. The air was startlingly clear. From other passes came more oxen, more bearded men. The wagons were beyond count on all the fingers of two hands. They went into scattered camps. The two Indian watchers in the Panamint Mountains saw the fires play for several nights. One night there was a particularly handsome blaze. One of the parties was burning its wagons. What was to ensue could be sensed, without much difficulty, by the Indian witnesses.

Early in the previous spring these people in the trough below had been part of that infinite host, drawn from every midwest and Atlantic town and hamlet, which had merged in the Missouri country, put forth across Nebraska and Wyoming, moved through the Rockies, and come out, a sprawling river of wagons, upon the intermountain basin. The procession of gold-seekers was of a length beyond calculation, and in its composite mind there was every shade of hope, fear, innocence and recklessness—innocence perhaps prevailing.

At Salt Lake City many of the companies had reformed. Among the California-bound were a certain number who were disposed to reject the direct trail. Straight ahead meant certain hardship. The longer way, around the Sierra Nevada by an indefinite southwest route, offered a bit of glamor. It was a route unknown. It might be rather pleasurable. In any event its waterholes would not be crowded.

Settlers about Salt Lake City supported their plans, with reservations. A few hardy Mormons had been over that course before. The travelers were advised to take a competent guide. Under his shepherding eye a happy outcome was predicted by way of San Bernardino Valley off the toe of the Snowy Range. Fivescore wagons were entered in the battalion. Half of these held women and children. With winter rains and first grass the hopeful pageant was off.

For a few weeks it enjoyed fair progress. Then disputes arose. Opinions hardened; near Mountain Meadows the train broke into two obstinate detachments. One, with eighty wagons, kept on behind its chartered leader to southern California and safety. The others, leaderless but bold, turned directly west.

There were twenty-seven wagons in this section. They pushed on for fifty-two days, practically without renewal of their food supplies.

The weather may have been rather delightful. It usually is, in desert country in the fall of the year. But to these travelers, reared in lands of deep grass and mellow forest, all was appalling. The intermountain basin, distressing at the start, soon became wholly hateful. Distances between feeding plots grew greater. Finally water disappeared entirely. The draught animals plodded on, weak and wondering. The men plodded on, the women and children rode, hungry and wondering too. One morning a bachelor member with a small wagon and a single yoke of oxen gave up the fight. He was buried at the trailside, and if anyone carved a headboard, none later could recall the name.

In spite of growing hardships, some of the party kept up diaries. "The boys found some rain-water in a basin in the rocks and took a good drink. West swallowed all he could hold and then told the boys to kill him, for he would never feel so good again." Five more days without water, and then a light fall of snow, caught in a blanket. In Forty Mile Canyon the Rev. Brier burned his wagons and packed everything on the oxen. "It was a fatal mistake, as we were about five hundred miles from Los Angeles and had only our feet to take us there."

An Indian cache of coyote melons found and raided; three oxen struck by arrows; one of the animals slain and eaten by the emigrants, its marrow a foul slime from emaciation—the tale grows somber. "Offered Brian Byron $5 for a biscuit but he refused. Old man Townshend left . . . One man laid down. Made coffee, went back and he was dead. Find body of Townshend; scalped."

Ultimately, and by now rather at odds with each other and widely scattered, they come to the region which is the eastern anteroom to Death Valley. The Valley itself is as yet unnamed, its existence unguessed. There are those in the caravan who think that perhaps, immediately beyond these desperately barren mountains, the gardenlands of California wait with fountains and garlands. Others are not so sure.

They are seriously thirsting, and there can be no more delay about feed and water. Just off the line of travel a small range of mountains sticks up into thin clear air. The crests look but a mile or two away. Even the most inexperienced know by this time that desert space is deceptive. But a few of the younger men set off to look in those folds for springs.

They come back with the report that they have found a mountain literally seamed and charged with shining silver. The report brings no cheers. What matters is that they have found no water.

The wagons creak forward. There are those who are sure, by this time, that they are beholding green meadows spangled with lakes and surrounded with leafy groves. They can hear the tinkle of waterfalls, the play of waves. They laugh a little wildly and point out these delights to friends who will not see.

A day and a night bring them to pools where the Amargosa River, which runs underground through southeastern Nevada, comes to the surface momentarily. But "amargosa" means "bitter," and these are brackish pockets. Travelers, mules and oxen halt gratefully, however, and drink deep. Illness follows. They are crossing, in this unhappy place, the eastern boundary of their land of illusion—California. Were their geography that accurate, the thought would bring them to laughter indeed. But there is a rude sort of salt grass here, so for a few days the party tarries. The stronger members accept the chance to go back and look once more at that mountain of silver.

A year later a member of this exploring contingent, Hugh McCormick, or McCormack, will be showing to other miners on the Coast what he declares is true booty from the mountain. Pounded out of its native rock, it is pure wire silver, as bright and shining as new-minted money. So abundantly did the metal fern and vein this mountain, the pilgrim will assert, that at a little distance the whole peak glistened with it; close inspection, he says, revealed the alp practically bound together with those white metallic strings and coils. But this is a story for the future. The party moves on from its bitter pools and enters the Valley to which its sufferings later give name.

As they descend they find themselves apparently getting higher and higher into new mountains. It begins to occur to them that the footings of these heights must be below the level of the sea. They come in time to the floor of the white sink. The beanpod-shaped depression is a day's foot-journey wide, a half-month's foot-journey long; its brown, red and yellow sides up-curve to a mile, two miles above the salty floor.

Now follows, though it is winter and the cool of the year, that outcome of innocence and folly which the Indian watchers could foresee. Starvation in the sight of food. Death by thirst while almost in sound of running water. Captain Culverwell, dropping out there on the flat with his face on his arm, to mummify under a desiccating sun. Pinney and Savage, "back-packing" across a northern curve of the valley and escaping, while their nine companions branch off on a short cut into tragic mystery. The man Fish dying in a gap of some hills beyond. Ischam, crawling to death on his hands and knees. Robinson and McGowan getting across this and other depressions, but succumbing at the base of farther-lying mountains of granite. Frank the Frenchman, wandering away to be found ten years later, living as an Indian among Indians. Mrs. Brier, leading two children and carrying the third, a nine-year-old boy, on her brave slight

"The sun swung up the far side . . . coloring the tip of Telescope Peak."

"There are those who are sure they are beholding
green meadows spangled with lakes . . ."

"The strong light was flooding salt-paved Panamint Valley."

"Mighty gorges offered them a variety of gateways."

back. Manly and Rogers setting out for the distant seacoast to fetch help and coming back with their mule, laden with food, weeks later. Mrs. Arcane and Mrs. Bennett, crawling out from beneath their wagons, weeping for joy at the succor, gathering their broods and ascending the heights at last—"Good-bye, Death Valley!"

The watching Indians, father and son, clinging to the flanks of the mountain they called Chiombe, could have told these entrapped travelers that the straight-podded honey mesquite almost always indicates where to dig for water; that their squaws for centuries have made flour of the screw-bean mesquite and of the little brown seeds of the spiny ironwood; that the beaver-tail cactus, after being rolled in sand to remove its vicious hairs, is succulent when boiled.

Lore such as this, the ancient heritage of folk who dwell in hard places, could save white bodies. But the Indians have never seen oxen or wagons before, or great-bearded men; so they keep aloof. Those intruders who die out there are beaten, in all likelihood, by plain stark fear of the region as much as by its rigors. But one thing does not die. This is the story of that mountain of silver.

THE mountain grows with each telling. It grows, and it moves hither and yon on the desert like a whirling sand-augur. The survivor McCormick, showing his nuggets to placer miners at Rough and Ready on the Yuba, warns that the find can never be of practical value—that it is located off in the middle of alkalilands hundreds of miles from any settlement. Moreover, he attests, no tools will dig this mountain unless a blacksmith stands by to sharpen them after each impact. His warnings merely add stature and brilliance to the pile.

Presently the treasure-mountain has a name. That unfortunate Towne or Townshend, tradition now states, gathered up some silver while on the long trek and hammered it into a front sight for his rifle. The so-called Gunsight Lode becomes a compelling magnet, and its existence as a tale—whether or not as a deposit—draws men by the scores into ravines and crannies of the shadeless country.

The beautiful thing about a lost mine is that the more thoroughly it is lost, the brighter it shines. And as Gunsight gleams on an ever-receding horizon, it takes on lustrous companions. Sometimes it merges with these legendary consorts. Sometimes it stands distinct. John Goller was one of the '49 party who stumbled at last to the adobe ranch house of hospitable Don José Salazar in San Fernando Valley in southern California. To Don José and all others who would listen, the ragged traveler told of a treasure hill. For a quarter century the good Goller, who shod horses and made good, substantial wagons and stagecoaches for a living, repeated his story, and between bouts at the bellows spent his time and savings on expeditions that went off into the unrewarding desert.

By now the mountain was thoroughly lost, and not only lost but turned to gold. John Goller had nuggets to prove it, and recalled that at the moment of discovery he could have loaded a packmule with pretty samples, had a mule remained uneaten.

But the Sierra Nevada is a formidable barrier. For ten years the men of Eldorado found enough to do ransacking its western side. There the Mother Lode trails were alive with movement,—with ceaseless rushing to the strike of strikes that was forever just over the next blue crest. Shoreward called Trinidad Beach, where gold was to be shoveled up and pushed away in wheelbarrows. Then Fraser River in the far fir country, where gold was commoner than pebbles. No tale was too incredible, for the Mother Lode itself was proof that the incredible could exist.

Then, one January day in '59, a group of ragtags just east of the thin, tall barrier range—men who found themselves on that drab side because they were too tired to cross, or too ornery to be received—sunk their picks into the hill known later as Mount Davidson.

They struck the top of a mine they call Ophir. Its outcropping of yellow-charged quartz bewildered them. But they laughed and got very drunk when some breathless newcomers bought them off for a few thousand dollars. The amused billies knew that beneath the surface glitter was a pesky blue-gray clay which sucked at the feet and stuck to everything, and they wanted no more of it. Let the purchasers try to shovel that. There is hilarity on both sides of the transaction. For the lode is the Comstock, and in the next two decades the gluey clay, which in reality is silver of staggering worth, will produce boom, statehood, panic, collapse, boom and reboom; it will produce Bill Stewart, John P. Jones, Mark Twain, Sharon, Ralston, Mills, Hayward, Baldwin, Hearst, Mackay, Flood, Sutro and Fair; it will produce Virginia City, Gold Hill, Silver City, and a fabulously reborn San Francisco; it will yield speculation, splendor, demoralization, ruin, and all the rowdy fandango that can accompany the ripping of $350,000,000 from a single hill.

The most grotesque of dreams, then, can turn out to be real. Nothing on top of earth or under it was ever more fantastic than the actual gold-and-silver Comstock. Earliest assays along its surface were sufficient to create a mad rush eastward from California, a rush over mountain roads fairly paved with capsized wagons and gear.

Silver-hunting was new to the West, but it became stylish as the Comstock revealed its stores. Any shrub or stone in the sunlands might conceal its duplicate. Men set to scouring each mile of the intermountain basin. With the new enthusiasm for silver, prospectors again gave thought to the forgotten corner that lay off the southeast toe of the Sierras—the region of McCormick's mountain of silver, Towne's Gunsight vein, and—while they were about it—Goller's gold-stuffed peak. Chiefly it was the Gunsight that drew them. That had become the general title for all lost leads.

Among parties organizing anew for the Gunsight hunt, fifteen persons headed by a Dr. Darwin French started from Oroville and turned their horses adown the long interior valley. They crossed the Sierras where that wall breaks down at its southern end and picked their way over a succession of farther-lying ranges.

The Gunsight by now had taken on trimmings. It occupied a region where Indians shot golden bullets. The French party did not find these well-endowed Indians, but in regions where the aboriginals lived on pine nuts and grasshoppers they found a little golden quartz. One of these locations became the camp known as Coso, meaning "fire," and a superheated location it was, in a world of slag mountains and baked gravel.

To the east was another bare range, the Argus Mountains. Across an alkaline trough later known as Panamint Valley rose more heights, rugged and steep, culminating in a notable peak. These heights were the western wall of Death Valley. John Lillard of the French party named them the Panamints for their handful of Indians. Where the Indians got the name is unrevealed. The explorers crossed the Panamint Mountains and went down into the kiln beyond, where at a spring against the base of the high peak they beheld, still bright and unrusted, wagon chains and utensils as left by the Forty-niners.

The Gunsight still was missing though the French party searched well for it. In the fall of 1860, a certain S. G. George and party took up the decade-old hunt. The group of four included William T. Henderson, who had helped run down Eldorado's arch-bandit, Joaquín Murieta, and whose hand had severed that chieftain's head for identification. In the seven years since elapsed, Joaquín's head had floated in a jar of alcohol in a San Francisco resort; and Henderson when alone was said to feel pursued by a pistoleer in a Spanish saddle who would canter up beside him and demand his head back. To a man so haunted even when sober, the region Henderson now entered was not soothing.

The quartet followed Dr. French's route into Death Valley, and on its way out came to a long canyon running down the west side of the Panamint Mountains. Here grass and good water invited. Camp was made under a wild rose thicket growing beside a spring. From this pleasant base the whole region was prospected. Above rose the commanding peak whose feet stood below sea level on the Death Valley side and whose head, more than two miles aloft, was necklaced in pines and capped in snow. Henderson ascended this eminence, perhaps to get away from that headless horseman.

What he saw from the top was strictly unusual. Range upon range of desperate mountains, separated by sandy scoops, stretched away on two sides. Immediately beneath on the east was the beanpod-shaped depression whose very appearance was calculated to scare wits out of a man. On the other side was the smaller but equally austere Panamint Valley. Against the west, radiantly clear, sprang the Sierra Nevada.

It was December when Henderson looked across lesser obstacles to the overwhelming bulk of Mount Whitney and its neighbors, which were white for several thousands of feet down from their crests. He was standing directly between the highest and lowest spots in the United States. They were separated by about seventy-five airline miles. Southward rose the Calico Mountains and, faintly seen across the Mohave sandplain, the far San Bernardinos concealing coastwise valleys. To his north lay the Grapevines; east across Death Valley were the Funerals, their slopes festooned with gray and black slate. Henderson's summit was the steeple to a mountainy rooftree. It was gouged on both sides by stupendous canyons ripped open and polished by past deluges. It was all a raw, jolting panorama. Considerably impressed, and unaware of its Indian name, Henderson called his coign Telescope Peak.

The solitary explorer went down to his camp and reported what he had seen. He especially mentioned the canyons gashing the range. They might contain anything. Treasure often lurked in the most unlovely spots.

The prospectors moved on, and fell in with William Alvord and party of seventeen. These were treasure-hunters also, and were looking for a cliff reputedly shimmering with metal which Alvord once had glimpsed up one of the deep ravines south of Telescope. This was directly behind and through the granite wall from the spot where the Arcane-Bennett pioneers had camped and suffered in '49. It was off the classic track of the Gunsight, but if Alvord's story was true his find would do. Alvord presently gave up the quest, but a few days later Henderson's partner, George, picked up an Indian for guide and sought it on his own.

This Indian was a young buck of about twenty, slender and tough as the great-leafed desert willow, who as a boy eleven years before had crouched behind a rock and watched those great-whiskered men invade Death Valley. He had ceased to fear the white men, for one of them, after chasing and capturing him near Emigrant Spring, had let him go undecapitated. In course of time this Indian was to become quite a chief, the builder of coaching roads to the rousing camp of Panamint, and in the end the possessor of its abandoned mules and stagewagons. He was to live perhaps a hundred years, and to see the age before ox-wagons melt into the age of the motor car and the airplane. But just now he was a puzzled stripling, not at all sure what George wanted of him, but the possessor of a secret which the prospector had come miles to win.

The spot to which the Indian led the white man was tumultuous.

Slanting upward from Panamint Valley was a broad alluvial fan or ramp of gravel, the debris of past cloudbursts. Terrific must have been the force that had tumbled all that stone out into the valley. Five hundred feet vertically higher, George and his guide entered the yawning mouth of the gorge which Alvord had alluded to, calling it Surprise Canyon. Cliffs and slopes were completely bare. Their surface material had long since been hurled down sunbaked slopes. Ahead, though curtained by the windings of the canyon, was a collecting basin of some thirty square miles whose only outlet for rain-water was this tortuous funnel.

The walls closed in. The sky became a thin blue string. For several hundred yards the passage was scarcely wider than a man's outstretched arms. Down through it tumbled a snow-fed brook. Several miles of rough going by a game trail brought the two humans to an abrupt cliff over which the flood-stream yelled and bounced.

It was a beautiful cul-de-sac for a murder and George concluded he had seen enough. Every gesture of the Indian guide took on double meaning. Putting the savage in front of him, George hurried him down the gully. He had not seen the promised treasure, nor had he reached a certain amphitheater, six thousand feet above the canyon's mouth, which under the name of Surprise Valley was to be the town-site of future Panamint. But he had seen cliffs and gorges enough to fill his dreams for months.

On the homeward journey the George party again picked up Alvord and a companion. The pair were separated from the rest of their friends and Alvord was sick; his companion was holding him on a burro and the two had been living on jackrabbits. George gave them supplies and saw them safely out of the desert.

Alvord, after this gainless experience, made yet a third trip. He was obsessed by that something which he had once seen in Surprise Canyon's upper reaches. This time he took along a partner named Jackson, and he must have promised Jackson a gorgeous eyeful. For when the canyon became steep and wearing on the nerves, and no Gunsight had yet appeared in the midst of it, Jackson in his great disappointment rose and slew Alvord. This seems to have been the first of those numerous homicides which were to decorate the history of Surprise Canyon.

S. G. George likewise was not done with the locality.

In 1861 he returned with friends. They proceeded to Rose Canyon, that previous camp ground in the Panamints, where signs of antimony and silver led them to organize the Telescope Mining District. Into the flank of the towering mountain they drove a tunnel for 125 feet; whereupon Panamint natives, disliking this invasion of their meager hunting grounds, crept down and despatched the four men who had been left to carry on the work. The tunnel became a grave, and silence descended over the Panamints that endured for some years.

But events were shaping. One day in 1862 Bob Haslam, a famous courier, was streaking across central Nevada with the hurry-up mail matter of the Pony Express. He was clacketing down a steep gorge in the Reese River country. His horse, said the tale which reached men on the Comstock, kicked a rock askew. In spite of Indians at his heels and the prod of the hardiest express schedule ever devised, Pony Bob noticed something about that rock. It glinted with silver. Bob took it to Virginia City and the rush for Reese River was on.

Three men, a Scotchman, an Irishman and a Dutchman, joined plans and started from Los Angeles for Austin, the camp created by Pony Bob's discovery. Messrs. McLeod, O'Bannion and Breyfogle had more energy than wisdom. There were four hundred miles to traverse—every mile a mile—and they determined to lay a course straight for their mark. They left San Fernando Mission, that haven of early-timers, and climbed up to the Mohave sandwaste; they swung around the Inyo ranges and across glistering soda valleys; they ascended a particularly lofty range, probably the Panamints, from which the view on the first of June can be sufficiently disheartening. Death Valley lay below and it

gleamed with the gleam of hot horrors. Yonder rose other
sterile ranges. What matter? Fortune beckoned.

In a "tank" or shallow basin in the rocks the three found
water. The ground was rough and uneven and Jacob Breyfogle, searching for a flat spot, cast down his blankets apart
from the rest. In the night the Indians came. Breyfogle
heard two mortal yells. That was enough. Grabbing up his
shoes, falling to his knees and struggling up again, he raced,
ran, stumbled. By morning he was down at the edge of the
sink they had viewed the day before. He filled his shoes with
alkaline water and lugged them for canteens. More or less
carrying his blistered feet in his hands, he came that night to
the mountains opposite.

With morning, his searching eyes detected a moist-looking
spot. Climbing for the fresh drink that he now must have or
perish, he kicked up "float."

The chunk was grayish rock fairly aswim with gold.
Higher, Breyfogle found the out-crop which was the source
of his bright chunk. It was regal.

The ensuing days were days of madness for this man who
had only brackish water and raw mesquite beans to subsist
on. But he pushed two hundred and fifty miles, straight
across the Amargosa and farther-lying scoops and rises; he
found refreshment at Baxter Spring and tarried until his
strength returned; he crossed Smoky Valley nearing his
goal; and there he was picked up by a ranchman. He was
carrying his bandana. It was full of gold ore. But on returning with new allies he couldn't find a trace of that thumping
ten-strike. Such was one version of the "crazy Dutchman"
story. Another version had him emerging from nowhere with
a bag of nuggets which he said he had found clutched in the
hands of a dead prospector.

The Breyfogle tale gave another lift to the Gunsight search, and led to important discoveries. On a hill 8,000 feet above the ocean and 4,000 feet above a nearby dead sea known as Owens Lake a strike was made of ore averaging twenty per cent lead and $60 in silver to the ton. There was no timber of any kind at hand and water had to be fetched ten miles; nevertheless Cerro Gordo developed and so flourished that by 1872 it was producing $1,500,000 a year, shipping its bullion down to Los Angeles in heavy bars on big freight wagons.

Meanwhile mighty Comstock went on providing. San Francisco, which had first won its importance as the entry port to the early-day gold diggings, now was living by and for the ungrudging silver lode across the Sierras. It was rearing theaters, hotels and mansions on the reality and the promise of Virginia City's output. The world of finance hung on the antics of its Comstock stock boards.

By 1872 that miraculous Nevada lode had sent to the surface $120,000,000 in silver and $80,000,000 in gold. It had produced jubilation, heartache, $10,000,000 worth of litigation, and in its William M. Stewart at least one lawyer-millionaire. It had produced fires, cave-ins, heroisms, and in wide-bearded John P. Jones at least one multimillionaire mine-superintendent. It had summoned from the depths of the Crown Point mine, at an opportune time, $9,900,000 when all else seemed played out and had elevated Superintendent Jones to the United States Senate to sit beside Bill Stewart. Now, on that same lode, picks and crowbars were ringing close to another nugget as long and wide and twice as high as the national capitol at Washington—that ore-body charged with eighteen hundred tons of silver and eighty tons of gold which was to be known as the Big Bonanza, and which was to upset the politics of nations.

This was the distant situation when, one winter evening, three prospectors, one of them close to insanity, staked out their jacks in sight of the Panamints and set about preparations for a forlorn dry supper.

# 3

## SURPRISE VALLEY

A man knelt in the lee of a bank, fanning a small fire. His wide frayed hat was a veteran of many sandstorms, and plainly would fan many a campfire more. Above the barranca's rim the sharp evening air was in motion. The sun was down behind the whitey blue Sierras. With the blaze going to his satisfaction, the man sat back and lighted up his pipe.

Another figure stirred in the burro-weed, sat up, shivered.

"How you feeling now, son?" The man at the fire eyed him judgmatically. "Still of a mind to take a pot shot at somebody?"

"I can't talk." The words were rasped.

"Don't try to, then. Better lie still. Friend Dick'll be back before long, maybe with a couple o' rabbits. Christmas Eve dinner—and we'll make salad of the ears!"

"I couldn't eat," shuddered the sick man.

"So. Well, as to that"—the pipe puffed—"maybe we'll all be relieved o' tryin'. Dick's been gone an hour, and I ain't heard any gun."

Minutes passed. The tips of the Sierras purpled, their snowy shirt-fronts turning to dirty gray. The fire-tender, hearing a note of new distress in his companion's breathing, put a kettle on the coals and poured in some mouthfuls from a canteen, shielding his act from view. When the small store

of liquid began bubbling he added a handful of long-jointed
stems pulled up from the gravelly bank behind him.

"Drink this," he ordered. "It's got lots o' names. Mormon
tea, desert tea, squaw tea or teamster's tea. I call it prospector's tea, and many a dish I've had with a bit o' broiled
chuckawalla." He watched the drink go down. "That'll be
all, though, till we get to Coso."

"Where did you get the water?"

The veteran said nothing. It was his last small swallow,
hoarded from habit. He had seen the time when men flung
themselves to earth to dig with their fingers, as though they
would tear apart the dry mountains; or, stripping off boots
and clothing, dash with laughter into the shoulder-deep blue
of some mirage. "I've chased some mighty pretty lakes,
myself, that weren't there," he could recall.

There was a crunch on the plain above. "Dinner for three!"
announced someone, and flung down a solid haunch of flesh.

"Jackrabbits grow big here," grunted Bob. He cut off
some strips and set them on sticks about the coals. In a moment, "Here, young feller"—to the one bedded among the
pack saddles—"eat this, and stoke it away hearty. We've
got some regular miles to do, tomorrow."

The Sierras went black in the west. The sky became pow-
dery with stars. The three figures brooded in the light of
their campfire. Its fuel, dry creosote brush, would blaze up
and die down quickly.

"Here's your pepperbox, lad," at length said the man
who had brought in the meat. He tossed over a contrivance
of four bunched barrels. "Think you can keep it in your
pocket now when you see me around?"

"I'm sorry, Dick." The young man caught and put away
his weapon. "I guess I was pretty close to the line. That

trouble's over. It's wonderful what tea and rabbit can do for a man."

"Yep," adjudged Bob the cook, "but we'll have to put one critter's pack on our own backs tomorrow. That rabbit of Dick's shore had ears."

Tall and salty are the yarns, lovely the faces of women, deep and rich the treasure-lodes, deep and rich and homely the philosophy that can be evoked in the outdoor country at the end of a weary day. Of the four themes Bob Stewart selected philosophy.

"The desert isn't cruel to those who understand her. The plants bloom and go to seed here, same as everywhere else. The little beasts ain't hard to get along with. The fiercer they look, the blunter their teeth. Of course the sand gets hot, and at times a man would give something for a pail o' water. But there's comforts out here too. There's them mountains over yonder, changing all the time. There's what happens to these dry bushes along about March, after a brisk dash o' rain. *Then* you see a garden! And there's the smoke o' little camps. Friends, you'll travel the whole world to find perfume like a fire of ironwood, burning pleasantly in some dry wash, or the burnin' smell of the gray sagebrush, the kind that grows in the higher country. Aye, with five gallons o' water the desert most anywhere is mighty fine—if you don't camp in cloudburst country in July or August, or lay your ear on a niggerhead or cholla cactus."

Christmas morn dawned, as it had dawned eighteen hundred and seventy-two years before, on three men trudging over endless sands. Not camels, however, but shambling jacks bore the shovels and gear of this trio. The sun swung up the far side of Funeral Mountains and Panamints, coloring the tip of Telescope Peak and, as Bob Stewart knew, filling the trough between with garish light. Then the strong sunlight was flooding Death Valley's westerly companion, salt-paved Panamint Valley. It slid by turns up the Argus range and up the Cosos.

On the far side of the latter it found the three men slowly ascending. These were approaching those hot mud springs found by Dr. French a dozen years before. There were scores of such springs in the infernal place; some were in constant motion, boiling and bubbling. In one good-sized oblong basin the strong alum solution rose and fell a foot or two every few minutes with the regularity of a tide. The ground was hot. Steam issued from crevices. Nearby sulphur banks and surrounding walls of lava, pumice and obsidian showed that Pluto had only lately retired from this region. Nothing was fit to drink.

The three pushed doggedly on; they were making with some anxiety for the settlement of Coso, where a handful of Mexicans were still working over Dr. French's early-day claim.

Richard C. Jacobs, leading the party, had been twenty years on the mineral trail. Its windings had led him half the length of the Mother Lode, with a halt with pan and shovel at Volcanoville in the California placers. In his time, Jacobs had also been one of the legion of pony express riders who connected the isolated camps of the West one with another. He was now about forty-two. W. L. Kennedy, who had started from San Francisco with him on this expedition, was less the miner and more the man of commerce. East of Walker's Pass the two had fallen in with Robert L. Stewart, an old-school prospector of the Inyo Mountains, who had come to California in 1857, following the thirty-fifth parallel with one of the government's railroad surveys. Known to those who crossed him as a smiling, unemotional, cold-blooded fighter, and to those who needed befriending as a wise, generous, understanding Samaritan, rugged Bob Stewart had elected for himself a wandering life on the dry side of the Sierras. There he had cut out, for lifetime investigation, a chaotic parallelogram a hundred and fifty miles square. Cheerfully he had accepted the company of the two other wanderers on this portion of his unending search.

They reached Coso the following evening. There was little said to them but a "Bienvenidos, viajeros" when they cast themselves to the waterhole. The twelve-year residents had seen wayfarers arrive in that state before.

For some days the travelers rested. The Hispano-Californians went on feeding ore to their rude little mule-powered arrastras and their women went on baking their immemorial tortillas. Bob Stewart refreshed his understanding of the regions eastward. To all questions the Coso men shrugged, and spoke of "los bandidos" and "los feroces Indios." Inquiry about riches in those farther hills brought the answer "Los muertos no hablan."

"The dead do not talk," mused Stewart. "Well, we'll see what live men can find. Vámonos!"

The travelers pushed on, thrusting down and up. Between the numerous parallel ridges all was salt, soda, alum. Once more water became a problem.

Bob Stewart, watching carefully, at the mouth of one canyon came upon Indian picture-writing in a shallow cave. A troop of deer, all headed one way, paraded redly through the soot of forgotten hearths.

"Between here and the Sierras there's hundreds of writings, painted with red earth or nickel into the black rocks," he expounded. "I've studied them a lot, thinkin' some might lead to deposits of minerals. I've come to the conclusion they're mostly Piute, Shoshone or Chemehuevi calling cards, or challenges to a fight, or maybe the valentines of young bucks in springtime. But when you see a carved or painted herd o' deer, all pointin' one way, that's to water. Another sign is a little pile of stones, with a white one on top. And there she is."

So they came to a rain-supplied depression in the rocks, and here they refilled. Then they set forth along the footings of the guardian range which separated them from Death Valley.

Mighty gorges offered a variety of gateways. The explorers climbed up each of these and peered within. At the mouth of one, Jacobs picked up a piece of soapy-looking rock. Advancing, he came upon a companion piece.

He handed the chunks to Bob Stewart. They were a pale yellow-green, lying on the palm with a waxy luster, and they yielded to the imprint of a strong thumbnail. The cut so produced was shiny.

"Horn silver," pronounced the old-timers.

The stones were dissimilar from any under foot. Plainly they had been washed down from heights above. They were better-looking float than any Stewart had seen in fifteen years of probing.

The three men pushed up into the chute. They were all eyes now.

Perhaps other eyes also did some peering. It was ten years since Indian George had led the white man George, whose name he had since assumed, into this same canyon; since the white man had taken fright and turned downward with his guide pushed roughly in front of him. During the interval, other whites had come, usually in furtive fashion.

Indian George, who had grown tired of his ancestral diet of grasshoppers and coyote melons, had made friends with these newcomers. For work performed or favors rendered, he had received small doles of their beans and tobacco.

It is likely enough that some of those furtive white men also watched the trio ascend. To the pioneers it was important that the newcomers were not peace officers.

A winter-swollen stream charged rowdily down, on its way to death in the gravel far below. Bob Stewart, familiar with the sort of floods that could sweep such gorges without warning, was for traveling in light marching order, so the burros were turned into a bit of shrubbery and their packs went to human backs.

The three men occasionally picked up more float. It led them on and on. Lateral galleries opened off in a jumble of routes. The prospectors stuck to the curving staircase of the main stream, though their watery guide occasionally ducked underground. All was exceedingly steep, rough going; in five miles of forward progress the men mounted a full vertical mile, scrambling over frequent obstructions.

But by this time verdure had begun to stud the walls. Clematis appeared in the shaded nooks, clambering over stream-bank willows; various brakes and ferns; hardy wild roses; finally the velvet-gray piñon. Trees! The desert world was changing to alpine.

It was nearing evening, and this was a good place to halt. Bob Stewart set about making camp.

When the fire had provided coals, he carefully laid two pieces of float in the red heat. The prospectors' heads went together. In a thousand far-scattered places Stewart and Jacobs each had made this test, usually without luck. But this time the experienced pair turned and looked at each other. The stones had bloomed out in little metallic globules, silvery beads.

Kennedy meanwhile pushed ahead up the canyon, searching for firewood. In the closing dusk he pulled up short.

A man was there before him. The fellow was sitting slouched in a corner of the rocks. But the staring eyesockets were innocent of orbs and the limbs, bare of flesh, were in a posture of slumped and grotesque weariness. Had the fellow flung himself to this attitude after some kind of an eccentric dance? Yes, after some kind of a dance. A half-inch hole was blasted through the skull.

"Bob, do you really hold there's a Gunsight mine somewhere?"

Kennedy asked his question thoughtfully while the three companions sat around their campfire.

"Well, friend, most likely there is." Leathery Bob Stewart looked long into the coals. "Several people say they saw it, and many more have seen silver that those people carried away from it. When a good many folks have seen a thing with their eyes and felt it with their hands, I'm inclined to judge that it's there."

"Then why hasn't it been relocated?"

"Who knows it hasn't? Them Forty-niners were half out of their wits. I've knowed men to get like that, when the thirst grips 'em. I've knowed 'em to try to shoot their friends. Never mind, son—that's by and done! But let's consider the point you've raised.

"Those old-timers we're talking about had no maps. Likewise, there were no names to their mountains or valleys— nothing you could anchor to—no landmarks. Of course they had Death Valley. But before they reached it and gave name to it, they crossed the Amargosa Valley which looks mighty like it. And after they got out o' those big sinks they had Panamint Valley to face, where even the burro-weed don't do much flourishing. It's safe to say our old-timers came out o' this region bewildered. Dates and places had pretty well run into each other. Mountain peaks had moved about—no two survivors could agree on anything. All they brought out was ore and half-crazed mental pictures.

"But there's been a hundred strikes in this country in the quarter century since, and some good camps started. Cerro Gordo may be the blessed Gunsight, up above the big dead lake. Lida Valley off in the north might be the Gunsight. Down near Saratoga Spring to the southeast there used to be a silver mine and a mill, but the Indians killed everybody. That might be the Gunsight. In this general region between the Sierras and the far side of Death Valley there's twenty thousand square miles, and in every mile of it twenty thousand rocks and gullies, and any knob or gully *might* contain the Gunsight—or another lode like it. I ain't seen 'em all. But I'm not done looking."

"When does a prospector ever get done looking?"

"The feller you found leaning up against the rocks yonder —he's most likely given up future expeditions. Poor old Bill Alvord! I remember when him and a party named Jackson set out to project this-here way. Maybe—mind you I only say maybe—their big strike was around just one more bend. We'll know tomorrow!"

Morning was still shadowy when the three set on. Storm-twisted junipers began to appear on exposed cornices, and side ravines bristled with piñon pines. Tributary streams were lashing down these forested gullies, filling the air with chatter.

Six thousand feet above Panamint Valley, the floor of the

main gulch became U-shaped, widening out to about a hundred yards. The stream was now brawling through a carpet of snow.

The travelers halted in astonishment.

Soaring cliffs in front of them, hiding a swing of the canyon, were ribbed with a greenish blue luster. In and through the weathered façade of azurite and malachite was the hint of untold tons of copper-silver ore. The precipice on the left ran up four or five hundred feet—a colossal chunk of mineralized rock giving evidence of quartz veins six to twenty feet wide. Gripped in limestone tougher than the surrounding slate, the metal-charged ledges had gradually emerged under the scourings of time and stood boldly forth.

Most mines are found by an outcrop at the top. But here the prospector could stand at the bottom of his find and see it all.

The vale extended on for two miles more, forking near its box-like head. There the granite sprang straight up another fifteen hundred feet and was corniced at the top with deep snow. From that ridge, the adventurers suspected, a clamberer could look directly down into the campsite nearly two miles vertically below where the Forty-niners had made their long stay of suffering.

The men advanced wonderingly. They were as three Sinbads marching into a valley of diamonds. Then they discerned, by a cairn or two, that they were not the first to invade this amphitheater.

A few nights later, in a parallel canyon a few miles southward, Jacobs might have been discovered very busy. He had built a rude furnace and drawn out bottles and appliances from his prospector's pack.

In front of him were five small heaps of colored rocks.

Jacobs broke up a morsel of the first, laid it on charcoal, strewed it with carbonate of soda and held the whole near his fire. Then he blew through a little tube of curved brass— an assayer's blowpipe.

The strong flame curled about the sample, rendered it intensely hot. Presently a bead of metallic silver appeared. Nitric acid had no effect on the powdered ore. But ammonia dissolved it. "Cerargyrite—horn silver," Jacobs grunted. "Bob, you called it."

Methodically he worked the other samples. One, a dark chunk with a green surface, gave off sulphurous fumes before the blowpipe, and with fluxes produced beads that he set aside for further tests.

"Might be copper glance, might be sliver glance," he said, "probably both."

He pulverized some of the rock and dropped it into a bottle, pouring nitric acid on the particles. Unlike the horn silver ore, they dissolved, giving off red nitrous fumes and producing a green solution. A dash of ammonia turned the stuff blue.

"No doubt about copper. Tomorrow we'll separate out the silver. It will take careful assays to show what we've got, but I think—by the look of things—that we're rich. Going to be hard ore to reduce, though—probably will have to be shipped to Wales for final treatment. But if it pans out as it might, this is going to be a camp!"

On the first day of February, 1873, the handful of Panamint Indians who occupied the upper slopes of Telescope and its companion peaks, and a dozen or fifteen white men who clung to hideaways in those sequestered folds, discovered here and there some pieces of paper fluttering under rocks or nailed to trees. One was under a stone at Flowery Springs in Windy or Wild Rose Canyon four miles north of the peak.

Another was in Mormon Canyon a few miles south. One was impaled on a twig at the "rancheria" or headquarters of the local Indians high in a little nut-pine valley known as "Quita-poa," and still a fourth was out on the Death Valley floor in a tassel of greenery known as the "Mesquital." The bits of paper said:

NOTICE!

There will be held at the camp of R. C. Jacobs & Co., in Mormon Canyon at the southern end of the Panamint Range, on February 10, 1873, a

MINERS' MEETING,

for the purpose of organizing a new mining district, and forming laws to govern the same. All claim owners are respectfully invited to be present at said meeting.

R. C. Jacobs
W. L. Kennedy
R. Stewart

To this rendezvous came a picturesque crew, dropping down silently by Indian trails through the ravines and passes. Their hips were hung with guns and knives and their faces were lost in whiskery mattings and the shadows of wide-brimmed hats. Trousers tucked into boots, shirts once red or blue but now dirty rust, eyes faded with long scanning of shadeless horizons and voices rough from hard fare and disuse, they arrived one by one and took places around the leaping flames.

Cautious but willing to be reasonable, as became men who had been sitting long on mineral treasure which they had no power to move, they formed a motley ring. Daniel G. Tipton, who had been a successful politician or business man somewhere, departing for causes unknown and destined to remain in these hills for more years than any; Jack Dempsey, a rough-and-tumble leader and natural dictator, whose gambler's activities along Nevada post roads had been interrupted by "Wells Fargo trouble"; John Wilson, previous name Curran, an Irishman of education who had left White Pine in a hurry and Pioche still faster; Hank Gibbons, a tough hombre and a rough joker; Jim Scobie; one Parker who was a satellite of Dempsey's; one Copely and one Chunn—were the local inhabitants drawn into conclave.

Jacobs pointed out to them that they might have the greatest mining prospects in the world, but that without capital or transportation their treasures were so much country rock. He offered them assays and customers for their claims, to be followed by mills, trails, roads, and civic development.

On motion of Jacobs, seconded by a thundering chorus, it was "Resolved, that we miners of the Panamint Mountains, here assembled, do ordain and establish . . . the 'Panamint Mining District' "—its boundaries extending from the wild rose spring out into the middle of Death Valley, south to Mesquite Springs, westerly up and over the Panamints and out to the center of Slate Range Valley, and then north to a point out on Panamint Valley's salt flats.

The rectangle was approximately twenty miles along a side and the terrain it involved was surely the most varied of any mining district on earth—ranging in altitude from 11,045 feet above sea level to 266 feet below, and in cover from desert holly and the iodine bush to pines and nearly permanent snow.

To the tipping of red-likker bottles and the play of flames the uncouth gathering pledged adherence to the mining laws written by Senator Bill Stewart of Nevada and adopted by

Congress the previous year, granting each claimant a length of 1500 feet and a width of 600 feet on any vein; conveying with each mineral claim the right to 320 acres of timber; setting up the office of recorder; and establishing the head-quarters of the new district in that upper portion of Surprise Canyon where it became Surprise Valley, and where Jacobs and party had found the cairns of these predecessors.

Jim Scobie and Hank Gibbons proved to Bob Stewart that they were already the owners of what he called his "Stewart's Wonder" claim, but open-handedly granted him a partnership. John Copely convinced Jacobs that the "Jacobs' Wonder of the World" also had a prior finder, whereupon Jacobs merged him and made him superintendent.

The meeting broke up when the demijohns were empty. The flames died low, and Indians peering down from Chi-ombe, which white men called Telescope Peak, may have wondered at this ceremony which was both the beginning and the end of "law and order" in that hidden region.

# 4

## "GREATER THAN THE COMSTOCK!"

T<small>REASURE-STRIKE</small> news travels by mysterious telegraph. Word of these events passed from juniper to saltbush, from pancake cactus to rabbit brush. It reached the wide, white "lake" in the Slate Range forty miles southwest, where John and Dennis Searles of the old Darwin French party were building vats for boiling the stuff of the lake floor into what they suspected would be merchantable borax—a drug-store commodity just then in much request at twenty-five cents an ounce. The word mounted to Cerro Gordo two days off to the northwest, where, a mile and a half above the sea, trains of ore-laden jacks were constantly on the move and the black of furnace chimneys smudged the sky. It reached Havilah and Kernville, important gold camps in the Sierra Nevada.

From the crest of that range it sprang to Bakersfield, and another leap took it to San Francisco Bay.

San Francisco in this hour was not yet the ornate town it was shortly to become. Its many-bathroomed Palace Hotel and its bay-windowed Baldwin were as yet only dreams in the lively minds of Ralston and Lucky Baldwin, plungers and optimists extraordinary. Its Nob Hill, reached by ingenious cable car, was as yet unadorned by the many-storied piles of Comstock kings and railroad emperors.

But the town had its Russ, Lick, Cosmopolitan, Occidental and Grand—hostelries of suavity and distinction. It had its brilliant theaters and its Lotta Crabtree. It had three hundred and forty restaurants, conducted by discerning chefs. It had its hills, fogs, trade winds; its Bank of California; its Barbary Coast and its spar-lined shore.

The town had just been yanked out of a fearful depression. The slump had been caused by a disastrous falling-off in gold and silver production along the Comstock. The recent upward tug had also come, as usual, from that lode of infinite variety, the Comstock itself.

Comstock's hidden treasuries were strung like beads along a few miles of the breast of Mount Davidson, with intervals of barren or "country rock" between. Today, horn silver could practically be cut with a jackknife, and millions would be in sight in a single nugget. Tomorrow, the pick of the miner might strike on hard borrasca, and financial smash-up would be the lot of those who had margined their guesses.

It was hazarding over what each new day's picks would strike along the Nevada lode that made all San Franciscans gamblers and the streets paved with the pearl of heaven or the hot stones of hell. Anybody could place his bets, and did so, though he knew that insiders were cheating outsiders and men in the mines cheating both. Today, shareholders would consider the walls of their selected silver mine as well defined as the curbs of Montgomery Street, or point to a deposit so obvious and pure that the only wonder was that it had not been stamped into coin by nature. Then, quickly as tomorrow, the porphyry limit of this treasure would be reached and there would be more citizens in the dumps than ore.

These ups and downs were unceasing but exhilarating, and the capricious deposits of the Washoe hills were as truly the hope, the pride, the anxiety and the agony of San Franciscans as if they had been located in the middle of Market Street. The famous playground of fortune might elect its Senators for Nevada, but it paid its dividends or collected its assessments, sent its owners to suicide or wrapped them in ermine down there beside the Golden Gate. Just now the talk was all of the great Crown Point mine and its neighbor the Belcher, the biggest bonanzas to date in the history of the lode. Mackay and Fair and Flood and O'Brien might be the unglimpsed multimillionaires of tomorrow, but William Sharon, Alvinza Hayward and John P. Jones were the fortune's favorites of today.

Especially John P. Jones. He was everybody's hero. With his own hands and his own eyes he had gone down into the earth, gripping and seeing down there under Mount Davidson what no one else could feel or see. Twenty million dollars, in the bottom of a hole that a hundred experts had called worthless! John P. Jones had more than revivified the Comstock. He had set burning again the torch of universal hope, and had turned every man into a potential millionaire.

To this Comstock-dazzled city came the infinitely small tap-tap of the juniper-and-saltbush telegraph, murmuring of Jacobs, Kennedy and Bob Stewart's find in the Panamints.

The Panamints? Where were they?

A few persons looked them up. Six hundred miles away, off across all coastal mountains and several deserts. San Francisco, with a tried, proved Comstock only sixteen hours and 346 miles away by comfortable train, was not disposed to concern itself over ledges at twice that distance in fantastic hanging valleys hedged by deadly sandwastes.

Yet in the daily uproar that greeted the opening bell at

the three local stock exchanges, where at "call the roll" the crowds inside and on the street yelled like madmen in an asylum, one pair of ears picked up the word "Panamint" and listened.

E. P. Raines was a shabby operator with a dwelling-room on Stockton Street just above Chinatown and a "seat" on the curb outside whichever of the exchanges he happened at the moment to favor. He was a big man, of pretentious and assured presence, with ears cocked to every stock-marketing whisper, rumor or tip. Certain unfathomable activities for out-of-town clients kept him, if not in funds, at least in yesterday's collars and last week's shirts. Across his waistcoat hung a gold chain heavy with seals though a watch might often be lacking at the end of it.

Raines was, in short, a determined loiterer in the path of fortune. But fortune, so far, had been a taunting jade. The San Francisco Stock and Exchange Board on Pine Street, with its four angular stories and handsome cupola, did not know E. P. Raines within the circular rail where sat its members. But when the "caller" on his dais beneath a grand crimson canopy cried the names of the day's popular shares —as the brokers rushed for the platform to yell and gesticulate, and every client outside the rail and every visitor in the gallery rose to his feet—Raines was one of the hundreds who milled about outside, straining for quotations that to some would mean thousands in their pockets; to others tragedy; to others a drink and transient partnership in a double bed.

Raines watched men get suddenly rich, or pass to the Bay and its outgoing tide, but neither extreme of luck had selected him. His seat of operations chiefly remained the sidewalk in front of the Stock and Exchange Board. Here he read the *Alta*. Its huge pages usually contained small items of large interest, if you knew what to look for. There was the record of perpetual thievery, for instance, along the Comstock and all over the silverlands. The mills at Virginia City, said the *Alta*, were aware of a constant, exasperating loss in bullion. An organized gang seemed to be at work. Organization alone could do it, for precious metal no longer moved in the light form of gold dust. It was bulkier now, and moved as bullion. The bullion bars were heavy, and each was stamped with a serial number and the mark of its mill. The band evidently had a mill at some secret point, where the stolen bars were melted down and renumbered. Then it became the duty of some agent, acting at a distance, to receive the shipments and get them turned into coin at the Carson or San Francisco Mint.

The year 1868 had been particularly active along these lines. Andrew J. "Big Jack" Davis, a Virginia City gambler who looked like a minister,* was suspected of leadership. Occasionally his band turned from bullion thievery to highway robbery. One June night, Raines had perceived in his newspaper, the Overland Mail was moving slowly up Six Mile Canyon a few miles below Virginia City. Baldy Green, most skillful and unluckiest of drivers, was on the box. His passengers of mixed sexes were down inside, trustingly asleep. When three men sprang out, pointing shotguns, only one passenger saved anything. That was his private purse, which he hurriedly dropped down the back of his neck. The travelers were tumbled out and shaken down, the Wells Fargo box looted, and Baldy left standing in the road with only his whip to remind him of his professional

* Concerning whose palmy days and sudden taking-off, see *Treasure Express*.

status—for the robbers had stolen the whole coach and driven off with it.

There were three sacks of bullion of an assayed value of $3,584, Raines was made aware, in that stolen stagecoach.

Twenty-four hours later and two miles above the scene of Baldy Green's misfortune, an overland stage driven by Bill Blackmore was similarly greeted by desperadoes who sprang from the rocks. Being of no mind to walk in Baldy Green's pained footsteps, Blackmore laid on with his whip, and rocketed up the gully with his passengers jouncing and yelling but his Wells Fargo treasure intact.

Then, on the second day of August, came the detonation of large-scale highway work on the road out of Boise, Idaho.

It was the Portland-bound stage this time, the place was Pelican Station in the Blue Mountains, the interrupters were four men in masks and the objective was a $7,000 Wells Fargo shipment and something over $50,000 in charge of an army paymaster. In all, counting the forced contributions from passengers, the loot amounted to $64,000.

All this took place in very hot weather, Raines read, the thermometer at Boise hovering between 103° and 112°. Though August fog at the moment was blowing in through the Golden Gate and San Francisco was wrapped in midsummer chill, E. P. Raines mopped his brow as he considered that temperature near Boise—or, perhaps, as he thought of $64,000.

The gangmen were being pursued. Their capture was imminent. It was supposed they had buried their weighty loot. So far, so good. E. P. Raines knew a gold vein when he saw it, even in an item of small type in a newspaper as big as a bed sheet. He cut the paragraphs out and stowed them in his wallet.

But all this was five years before, and Raines had not advanced his post of operations beyond the sidewalk in front of the Stock and Exchange Board. Not that he was discouraged. Hadn't Lucky Baldwin, who owned ranches, mines and stocks worth millions, only the other day run a brick works, boarding house and livery stable; wasn't Ralston the Magnificent a former river-boat pilot; and had not that shrewd buccaneer James R. Keene, like Raines his admirer, been a recent impecunious schemer of the streets? In this growing country there was opportunity for any hustler.

So, somewhere in the two square blocks bounded by California, Pine, Sansome and Kearny Streets Raines picked up the rumor, born under Telescope Peak, that had crossed deserts and mountains. Somewhere in the same two-block area he found a friend named Vanderlief, who submitted to a touch.

Raines had often dreamed of his big chance. It found him a man of action. No sooner had he funds in his pockets than he was being ferried across the bay by the "El Capitan" and entraining at Oakland. Four hours later he was detraining at Lathrop in the San Joaquín Valley for connection with the southward-building railroad. With him went Vanderlief, carried on by Raines' enthusiasm and by a desire, perhaps, to keep an eye on his contributed grubstake.

There was a four-horse, very dusty stagecoach waiting at Delano, the end of the rails. This conveyed the two Panamint-bound explorers to Bakersfield, landing them at the French Hotel for breakfast. Bakersfield was the junction point for two stage lines. One ran over the Tehachapi Mountains to Los Angeles. The other branched off for Havilah, Kernville, and Inyo points east of the Sierras. Raines and his companion took east-Sierra passage.

Fifteen miles beyond, the long plain began to drop from under their coachwheels. Through foothill oaks and presently through pines a rude road serpentined. Occasionally it clung to steep slopes above sensational drop-offs. Quail and deer scooted on ahead. After several hours' alternate rising and dipping, but always curving, and with occasional grand views of the great valley spread out below, Walker's Basin was won. Here barley, oats and pasture grass grew tall in a field of black soil hemmed in by craggy mountains; here too the air grew bracing, and a mighty meal was provided. At Kernville on beyond—a lively camp in the embrace of the overwhelming mountains—transfer was made to still another stage which operated weekly to the Owens Valley settlements lying behind the rampart.

The way now wound along cliffs overhanging the south fork of Kern River, up through deep sand to Walker's Pass and suddenly adown long aisles of grotesque joshua trees. The desert, at this point, had marched straight up to the crest of the barrier mountains.

Telescope Peak, plainly visible, stood in the east above intervening ranges. Raines eyed it with emotion: would it be a second Mount Davidson? Was some knob or spur of it the Gunsight?

And indeed, why not?

The stage dropped down to the scattered stations along the base of the mountains. High above and a little to the north sprang fourteen-thousand-foot summits, the rooftree of the United States, cool and white under their unmelted spring snows. All was granite up there; no gold, no silver. Eastward all was desert, dipping and rising and studded with creosote bush.

No stagecoach wheels had yet pressed the rough terrain

that lay ahead of the travelers now, but saddle animals and a local guide were obtained at the station known as Little Lake. Natural "tanks" of several acres, let into volcanic rocks, cupped a pleasant supply of fresh water and were covered with myriad wild fowl.

Raines and friend set out into the sea of greasewood. The Sierra Nevada at their backs had been heavily forested; but the gaunt ranges ahead were largely capped and faced with dolomite, limestone, slate and eruptive rocks and there was nothing green. Buttresses and spires pushed up, forming everywhere a broken and castellated skyline.

Morning of the second day out from Little Lake found the companions and their guide early in saddle. In a pass they drew rein and sat a few moments. To their left and two thousand feet above, Maturango Peak culminated the Argus crestline. A mile below, still in steely gray, Panamint Valley awaited its bath of light. In summer it would be a crucible. On its far side, outlined in radiance, Chiombe or Telescope Peak dominated all. Ten thousand feet above the salty floor its head was snowy, and its outstretched arms flowed with rocky draperies. The silence probably contrasted strongly with the racket around Pine and Montgomery Streets where Raines was accustomed to do his mining. They rode down into and across the final sink.

At the head of the portal which opened to the riders and invited them up into the Panamints, Raines found the beginnings of a town. The pioneers of the place and a few newcomers were camped in caves or stockades thatched with boughs. Everything in sight had been monumented and the residents were waiting comfortably for the next turn of the wheel.

The location was clearly hazardous. Just what perils

were in store for the camp were indicated a few months later, when watchers might have beheld the lower-lying Slate Range on the southwest receive its first rain in five years. The skies darkened and that eccentric visitor, a desert "waterspout," wound its way down the black masses of the mountains. It moved across the borax lake in a zigzag course, struck the opposite hills, and burst its skin with fury against their tenantless slopes. In a twinkling the treeless gorges were choked with water, which raced down in incredible speed and volume, forcing huge boulders and spews of gravel out into the sink. In a quarter hour all was over and the sun had taken relentless command again; but the men living in the Panamints had obtained a fair glimpse of the awful power of nature's fire hoses.

The townsite of Panamint was otherwise well selected and was the only logical spot in the whole vicinity. The vale was high, cool and floored with bunch grass; the brook at its head rivaled the Sierras' freshets; the slopes held firewood—and the mountains held the apparent stuff of dreams.

Jacobs, Bob Stewart and Kennedy were resting on two great claims on the left. Jacobs with his crucibles and cupels had determined that the gray ore behind those green and blue facings was charged with sulphurets and chlorides of silver ranging from $800 to $2,500 to the ton. All was promising, heaven was kind, and the only thing needed was capital for forcing open the treasure-house doors. Raines, who could talk, convinced the proprietors that he was sent by providence to produce the capital. He wangled from Stewart an option on his Wonder without passing any cash, but promised him $20,000 before a pick should sink six inches into rock. He then turned his forensic skill on the other locators.

Jacobs sat tight.

A return trip to San Francisco developed no purchaser for Raines' options, the party he hoped to interest being out of town. But it produced journalistic attention in both Sacramento and the seaport, the *Mining and Scientific News* in September observing:

"A friend who returned this week from a trip to Panamint mines, furnishes us with some information concerning the new district, which is of interest . . . Some of the leads are described as cropping out from a quarter of a mile to two miles in length, and run from a 'knife-blade' up to thirty feet in width. . . .

"About one hundred claims have been recorded and a townsite has been located . . . The climate there is good, but very warm weather may be experienced on the road, which is said to be a rough one. The country is perfectly wild and unsettled, with no game, except a few mountain sheep."

Raines was not one to be permanently cast down by a cuff or two from fate. He managed to raise a second grubstake from the obliging Vanderlief and the pair returned to Panamint, where rapid and energetic talk extended that option.

First snows were just starting to fall when Raines, Vanderlief and Kennedy loaded three hundred pounds of rich rock on a couple of mules and set off anew down Surprise Canyon. They bent their course this time across the Mohave Desert and dropped to the southern coastside.

The trio with their bags of rocks reached Los Angeles on December 12th, 1873. The future metropolis of the Coast at this date was a town of about 10,000 inhabitants. Its adobe-walled, flat-roofed houses were set amid pleasant vineyards and groves of oranges, lemons, figs, olives and pomegranates. The atmosphere of Mexico still held the place in tranquil embrace, and its streets were paved, according to a correspondent in the contemporary press, with "old tin kettles, scrap-iron, shavings, cards, dead cats and horse manure." But already the American section had forward-looking merchants and a bustling chamber of commerce, and a brand-new horse-car line, which was much admired. A railroad had recently been constructed eighteen miles to Wilmington, whence steam launches and lighters conveyed passengers and goods to and from the sea-going vessels that dropped anchor in San Pedro harbor. Daily stagecoaches rattled off to Gilroy, a railhead south of San Francisco, and there were tri-weekly stages to San Diego and to Prescott, Arizona. There was also that stage line running over the Tehachapi Mountains to Bakersfield. But this was about the extent of modernity. Beyond the groves and vineyards all was flat dustland, terminating against the sea on one side and high mountains holding back high deserts on the other.

Raines, Kennedy and Vanderlief marched into the low-roofed adobe Clarendon Hotel, formerly the Bella Union, and spread their specimens on a billiard table. Messrs. Temple and Workman came in from their pioneer banking house next door and viewed with interest. The display quickly gathered a crowd.

This time Raines' program had been laid with care. He was not in the southern city for a purchaser. He was after a newspaper background and a wagon road.

Shrewdly he alluded to the location of those wealth-charged cliffs only two hundred miles from Los Angeles' doors. He reminded his audience of the important and growing bullion traffic of Cerro Gordo, Panamint's "neighbor" across a couple of mountain ridges. The output of the "fat hill" came down in freight contractor Remi Nadeau's ceaseless wagon trains, each prairie schooner followed by its back-action or tender and the whole drawn by a dozen or more mules. There were scores of wagons and several hundred mules engaged in the business. As the bars of lead-and-silver came down, for transfer by sea to San Francisco, lucrative purchases of barley, hay and merchandise went back. All this the Angelenos knew, but they were not averse to hearing it repeated.

Then Raines really launched into his subject.

With airy waves of the hand toward his samples he told them of Panamint's sky-raking ledges shot with copper-silver; of the limitless tons of this material in sight, and the vastly greater quantities that must be lurking in the unplumbed depths. When it was suggested that his eye-filling chunks had been plucked from a limestone setting and that limestone had no very high standing as a matrix for permanent silver deposits, he dwelt upon the Tirito and Almada mines near Alamos in Sonora, Mexico, where for generations a vein had been traced down through limestone of just such a character to a depth of 900 feet, in the course of which it had widened from three feet to fifty. So much for limestone as an enduring foundation for a mine; and anyway, Panamint's wealth was visible, not hidden in the earth, nature's chisels have laid those ledges open for depths of hundreds of feet.

Raines urged his spellbound listeners to go up there into Surprise Valley and verify his statements—an invitation that was purely rhetorical, in view of the difficulties of getting thither; and repeated that he had nothing for sale. This remark made the greatest impression of all. But given a 60-mile lateral for wheeled vehicles connecting Surprise

Canyon with the Cerro Gordo road, and Los Angeles would reap the benefit of the greatest tonnage of mining freight that ever lined the pocketbooks of a roaring western city.

So exhausted was Raines after each presentation of these facts to an ever-changing audience that he turned frequently to the bar. It embarrassed him greatly to find that he could not pay for drinks for all—nor even for himself. There had been some rascally hold-up men on the road out yonder —thank you, Stranger—and you, Mr. Reporter—but gentle-*men*, as soon as we get our freight teams and stagecoaches running, you will be repaid at the rate of a cask for every hooker, a barrel for each noggin. Greater than the Comstock, gentlemen, greater than the Com——. Thank you, Mr. Temple! And Mr. Workman! Please examine those specimens again. They will, as you perceive, mean much to Los Angeles if what's behind them is brought through here for shipment. Bigger than Virginia City will be this new camp, gentlemen, and every dollar of Panamint's commerce can be your own!

Los Angeles thought of San Francisco with her 188,000 inhabitants; her three and four-story façades; her eight horse-drawn car lines and that one line drawn up a sensational hill by a rope; her sidewalks of plank and cut stone; her streets insulated against the mud with cobblestones; her tall redwood houses with their extraordinary rash of bay windows—houses that were going up at the rate of a thousand a year; her brand new million-dollar Mint; her ships, quays, foundries, stock boards, cafes and temples of pleasure. All reared on foundations of Mother Lode gold and Comstock silver.

Mines, it would seem, could do rather well by a town that fostered them.

When Raines left for San Francisco a few days later on

the coaster "Mohango" he carried with him many column-inches of newspaper encomium, and left behind a well-started subscription list for a wagon road to Panamint.

Raines arrived in San Francisco on December 19th. This time the party he wanted was in town. Raines had four fast days in which to work.

SENATOR JONES TAKES CHIPS

# SENATOR JONES TAKES CHIPS

IN a rococo establishment distinguished for its long bar, its excellent food, and its liberal oil studies of the female figure, a closed door gave off at the rear. The gathering behind this oak panel was never large. It was clubby and exclusive; strangers were not unwelcome, but if they had only a few double eagles to toss to the banker they usually came out in a hurry with embarrassed grins. Drinks at the long bar were two bits each, never less. E. P. Raines had once or twice elevated its aristocratic glassware and, sipping slowly, had delicately cast eye over the luscious art on the walls and had less delicately sampled the lunch. But he had never, save in fancy, passed the tight oak door at the back.

But now it was a Raines with a purpose who entered the establishment. Salvin Collins, the chief proprietor, had just returned from the little rear room and Jim Wheeland, his partner and mixologist, was vigorously at work. Highballs had been ordered by the man who was dealing, and from the gold pieces that would be flipped to his tray Collins knew that the ex-sheriff of Trinity county and present Senator for Nevada would expect no change. Salvin Collins and Jim Wheeland could put real enthusiasm into preparing refreshments for J. P. Jones.

The ladies on the walls leered and lured, but Raines had attention only for the man in apron. He asked:

"The Senator is within?"

"He is, and a finer man you'll nowhere be finding."

"The table—is it full?"

"One gentleman, I believe, has lately run into borrasca and, finding the assessments heavy, is about to close himself out."

The ladies on the walls looked over their dimpling shoulders in petulance. Salvin Collins had picked up his tray and Raines was following.

The Senator was a bald, brown-bearded man of pleasant aspect. Beneath the neat black attire there was a hint of chest and arms of massive mold. He took his glass and waved the others around the table with a gesture that included the stranger and an empty seat. It was a gesture that seemed to roll back the oak wainscoting, the building's constricting walls and all surrounding edifices. The bald man's personality called for Space. One had the notion that his broadcloth irked and bound him, though he wore it quietly; that in an instant he could rise and split its seams, and with considerable success swing the sledge of Vulcan or the hammer of Thor, or with equal zest wield a muck stick or push an ore car.

To take chips in Collins & Wheeland's back room you needed something more than Confederate greenbacks. The Senator's look that swept the newcomer was inquiring.

What Raines sent sliding across the green felt was neither white mintage nor yellow. It was a chunk of dark rock big as a fist, that had lately been roasted over coals and plunged into water. Its surface was spangled with metallic blisters.

The Senator grasped it in a broad hand, turned it over under the shaded lamplight, pressed it with his thumbnail, and sent his level glance again across the table.

"Name's Raines. Just in from Panamint. That's near Death Valley. There's tons where that bit came from. Of course, if you're asking for coined money with the goddess of liberty stamped on each piece—"

Without comment, the Senator advanced to the intruder a four-color stack of chips. He admired banter and he admired boldness. Life is full of things to admire when you are forty-five with rousing good health, incalculable wealth, a sense of humor and the wide regard of your fellow men.

John Percival Jones came by it naturally, for he was himself a rock of a man and the rock was all ore. Son of a Welsh marble-cutter, there were generations of delving and stone-hewing behind him. As a babe he had been brought from a Herefordshire village to the howling wilderness of Ohio, where his parents had set about raising thirteen children. When Pacific Coast gold stampeded the nation he set out across Lake Erie with his brother Harry and other youths, made the descent of the St. Lawrence and tossed around the Horn in the fragile bark "Eureka," and arrived at San Francisco in the spring of 1850. Somewhere on his way up the west coast he passed or was passed by a little steam propeller containing among its crowded hundreds a certain young Bill Stewart, who was to share with him many a subsequent bed of adventure.

The Jones boys brought strong backs to the diggin's, and hands that were used to ax and shovel. On Feather River and Yuba, on Stanislaus and Trinity they scraped and panned with the other thousands. Mainly engaged around Weaverville, where life was thoroughly rugged and frequently

Virginia City during the Big Bonanza.

short, John P. Jones flashed qualities that made him justice of the peace, deputy sheriff, and through early war years the Union Democratic sheriff of the mountain county.

In that day and spot a magistrate or a sheriff needed common law not so much as common courage. The Jones kind sufficed. In '63 he turned Republican, stumped the camps, and landed in the legislature. Four years later as running mate to a less popular man he was licked for lieutenant-governor. Throughout his political ups and downs, then as in after life, he remained a man of the mines. He was deep-delving on the Calaveras, philosophizing the loss of his first of several fortunes, when a friend named Alvinza Hayward sent him over the mountains to the Washoes to superintend the Crown Point, a so-so property.

The Crown Point mine was between the Kentuck and the Yellow Jacket, Comstock favorites. Its 1100-foot shaft served all three by galleries at each hundred-foot level. The square-set timbering set up to support the great ore-vaults filled all drifts and stopes—a towering frame staging, brittle and inflammable, through which ran foot-walks never illuminated save by tallow drop or brass-ringed lantern. On April 7th, 1869, the day shift had just gone below when a foreman noted the odor of smoke as his car passed the 700-foot level. A hundred feet lower he found terror and confusion.

Men were running and choking, gas was swirling through the passages, and in an instant the cage was filled beyond capacity with fighting, clawing humans. Others leaped to dangle from its braces underneath. The cage made that trip up in twenty-five seconds. Men who tried to escape by climbing the hundreds of feet of vertical ladders were licked off by flame. Crowds around the hoisting works at the surface heard, deep beneath, an explosion that lifted the cage at the top two feet. During the next three days J. P. Jones was everywhere, heading attempt after attempt to reach the trapped men and later to quench the fire.

The heat was so intense that rocks below were found warm two and three years afterward. On the third day a last effort was made to get hose lines into the 800-foot level. Men gasped and crumpled in a few minutes' exposure to the belching smoke and steam. Jones sent them aloft, stuck to the task below with a single helper. A bulwark had to be smashed. While the volunteer held the light he gripped a sledge and was driving the last blows that would let in the stream of water when the man with the candle cried "Come away! Come away!"

"Two or three more smashes will do it," gasped the fading Hercules. Then he found himself in blackness. His companion had fled. Jones groped toward the shaft, finding his shuffling way by memory through a chaos of timbers. A single over-step would send him into the 300-foot pit. Strength and will nearly gone, he was crawling on hands and knees when his hands encountered a prostrate form. Jones dragged it to the cage and pulled the rope. He reached the top in a state of collapse with his thick arm still around the fellow who had abandoned him.

Always a popular figure on the Comstock, after that great fight John P. Jones found himself a champion among several hundred heroes.

Forty-five men died in the underground holocaust. Sealed tight with planks and earth, Crown Point and its connecting neighbors were left to burn themselves out. When Crown Point's shaft was reopened, nothing was produced for months save a steady stream of assessments. No dividend had been paid for several previous years. The stock now sank to $2 a share. Hayward, William Sharon and the Bank of California ring, who had their hands on everything on the Comstock, seriously considered ordering the famous mine's abandonment.

It was seven months after the terrific fire that Jones paid a surreptitious call on his old backer Alvinza Hayward.

Crown Point had sunk to a total market valuation of but $24,000. But Jones had found something too good for the Sharon crowd to know. He induced Hayward to buy all the Crown Point he could pick up and stake him to a part of it. Hayward acquired five thousand shares at an average of $5 before the other manipulators knew what he was doing. Sharon, who had been ruling the Washoe hills for years, was moved to jovian laughter. Since Hayward wanted the worthless stuff, he would fling enough at him to break his back. He began unloading. Hayward continued buying. The stock shot up. Sharon threw 4,100 shares on the market, glad to get a fancy price, and Hayward drew a check for $1,400,000 and took control.

Hayward's calmness shook Sharon to the roots of his being. For then the reason came out. Deep-burrowing Jones had uncovered a body of solid ore 200 feet long, and the $24,000 mine promptly soared to a market value of $22,000,000. On the third anniversary of the fire, J. P. Jones, the smoke-eating superintendent who never abandoned a comrade, was one of the richest men in the country.

If Jones had not lacked friends in his previous downs and ups, he had no lack of them now. They would have given him Carson Sink for a moustache cup and Mount Davidson for a charm to wear on his vest. He selected instead the Republican nomination for the Senate.

Here again he was squarely across Sharon's path. Sharon,

resident agent for the Bank Ring, an individual with un-limited capacity for getting what he wanted, had lately cast his own eye on the toga. The ensuing political shindy was carried on with all the rich enthusiasm of a barroom battle with no holds barred and the finish, with whoops and shouts, out across the sidewalk and down in the gutter.

In the course of the campaign, a gambler called on the Crown Point hero and proceeded to business.

"J. P., I control a thousand votes in this election. It will take about ten dollars a head."

"Who are these gentlemen that desire to sell their votes?"

"J. P., they are quiet fellows who will never say a word. It has taken me a lot of work to get them organized. But I've got 'em in my pocket, and I'll vote the lot for you for $10,000."

Jones sought further light.

The sport's voice dropped to a whisper. "J. P., they're up in the burying-ground. I've been a month cleaning the dust off those headboards and copying their names. But I've tended to the job in style, and every man-jack is on the great register."

Jones shook off this offer and the sport retired, warning him belligerently that Boot Hill would rise en masse and vote for his opponent.

Equally to the point was an Americanized Englishman who, knowing Jones' British blood, offered him the support of several hundred fellow-countrymen who wanted to organize a "Jones Britannia Club," but had no hall to meet in.

"All right," said Jones, "find a hall, pay the rent for six months and hire a band. See that the boys have refreshments and a good time."

At noon on election day the organizer hunted up Jones

and told him that the Jones Britannia Club hadn't voted yet and wouldn't until Jones paid a large assessment they had levied.

"I'll be right over," said Jones.

He mounted the little rostrum in the hall. "My fellow miners," he said, "I have taken a great interest in your club from the beginning. When I see English-born men come to America, and see them take on the solemn responsibilities of citizenship, I rejoice and say to myself 'These are worthy descendants of those men who set the Anglo-Saxon race free. They will be true to the land of their adoption.'

"These thoughts prompted me to help you in the formation of your club. A hundred years ago, when England had unworthy citizens she either transported or hanged them. I am satisfied that if this were still the custom, not one of you would ever have paid your own passage money. And now, asking your pardon for detaining you so long, I want to explain that my only reason for coming here was to have the pleasure of telling every last mother's son of you to go to hell."

To the wildcat yell which greeted this invitation he smiled tranquilly, waited for further developments, and then walked out. The Jones Britannia Club adjourned and rushed for the polls, voting overwhelmingly for "J. P."

Sharon charged his adversary with having deliberately set the Yellow Jacket fire in order to bear the market. That charge was too fantastic and Jones romped off with the legislature. Then, although election was certain, the people's champion scattered a half million in campaign expenditures "just to set a pace," he explained, "for the next man who runs." In these purer modern times such tactics might be

called corruption of the populace. The men of the silverlands had a better understanding of it. What was a nomination, if a man with plenty of money couldn't celebrate by buying his friends a few drinks? If the bill seemed large, why, so
were the number of his friends and the personal capacity of
each.

The victor went to Washington in the spring of '73. When
he took his seat in the Senate, much fun was anticipated at
the expense of the "metallic accident." Plans were laid to
heckle him out of countenance when he should attempt his
maiden speech. The attempt proved to be about as comforting
as taking a seat on a desert cactus. Whether bellowing down
a mine-shaft or addressing a marbled roomful of bearded
politicians, Jones turned out to be the master of a close-clipped,
homely style which conveyed remarkably clear thought with
convincing skill. Just about the time R. C. Jacobs and the
earlier locators in Death Valley's hills were organizing Panamint District, Washington correspondents were notifying their
newspapers:

"Jones has been here about a week, and is already a favorite
with his associates. His off-hand geniality and western 'bon-
homie,' backed up by a florid face, a restless black eye and
a frank, yet by no means rough manner of address, will secure
friends fast enough,"—as indeed they did; Washington hailed
him as a Croesus with more ciphers to his fortune than a
freight train had cars, and learned to love him as a liberal
host and a citizen of breadth. A widower without ties and with
means, personableness and a large good humor, he was the
catch of the capital.

By the end of '73, when Raines faced him there across the
green cloth at San Francisco, Jones' wealth was estimated at
from $5,000,000 to $20,000,000—an incalculable sum for the

day. Though he was distributing his income over ranches, mines, mills, horses, and the never-ending line of palm-extending friends, his efforts to keep pace with his profits were unavailing. The properties he acquired and the down-and-outers he grubstaked kept turning in steady dividends. Bystanders marveled at his luck. Gradually it dawned on them that this stocky, large-headed man with the straightforward gaze read men as he read quartz, and often discovered values at unguessed depths. And he was accustomed to placing his bets swiftly, whether on man or quartz.

So when the game broke up this night in the back room at 329 Montgomery Street, Raines had a tentative backer and J. P. Jones a $1,000 interest in some far-away ledges near Death Valley.

Here, for the moment, our story of the career of a western boomlet runs into borrasca. Raines seems to have spent this advance not in furthering the civic interests of Panamint but in crowding the excitement of several lifetimes into the next few hours. So Jones had misread his man after all, if indeed he had ever been fooled. Had he likewise misread his man's rock? When Raines woke up he was in jail, the ladies on Salvin Collins' public-room wall were presumably still laughing, and Jones had departed. The Senator was off for Christmas at his old home on the Comstock and his quarters at Washington.

Panamint's twenty citizens waited in snow and oblivion. Raines went sadly back to his post on the curb outside the San Francisco Stock and Exchange Board. As if to jeer him, Comstock was acting with extraordinary vigor. Crown Point's dividends of $5,300,000 for the year, and similar large sums pouring out of neighboring Belcher, not only had stimulated a new stock-gambling surge but were giving much encouragement to four schemers who had on foot a tremendous affair of their own.

Two of these men, Messrs. Flood and O'Brien, kept— and kept well—the Auction Lunch saloon in San Francisco's brokerage district. It was a place where men of affairs could lounge and gossip with confidence in the liquor and confidence in the sealed lips if not sealed ears of the bartenders. To this resort of gleaming walnut and polished crystal had drifted smooth-shaven, thick-moustached, hard-handed John Mackay and point-bearded, horizontal-moustached Jim Fair of the keen eye for the ladies and the keen nose for ore.

The two, who were capable miners from Virginia City, had an enterprising theory and a strong desire to talk to someone about it. The two hosts behind the bar had had long experience as good listeners and they fell into the familiar pose. To their surprise, what they heard sounded like hard sense. Comstock had proved its riches at many points along the five-mile fissure, but had provided no explanation for the barren spots between. It was inconceivable that the lode was not continuous. In the visitors' opinion it simply took a deep subterranean dive to unknown levels. What Mackay and Fair wanted was enough capital to buy Consolidated Virginia and other little-valued holdings between the long-time favorites Ophir and Gould & Curry, and to go straight to the bottom of all that rich probability though it took them to the center of the earth.

Fair and Mackay between them had $140,000. Flood and O'Brien had $60,000 more. The quartet formed a firm and pooled their resources. For $80,000 they picked up control of the Consolidated Virginia, the mine that would some day have a value of $75,000,000, and started after the adjoining mine known as the California.

Rulers of the Comstock thought they were crazy. But San Francisco had seen men equally demented come up out of a hole in the ground before this, triumphantly hugging millions. In the current year a drift had just been driven by the four-man firm from Gould & Curry through intervening property into Con Virginia at the 1200-foot level, following an often-lost film of blue ore. For the most part the seam had been but a string, often narrowing to a merest thread, always to be picked up once more by the unerring eye of Jim Fair.

Con Virginia had so far sent to the surface but two-thirds of a million, but men were beginning to sense that Jim Fair had hold of the leading-cord of fortune and the public was watching his course. A hundred million dollars, a European prince for his daughter, telegraphs and cables for Mackay, palaces and mistresses for Flood and unlimited whisky for O'Brien were tied to the hidden end of that labyrinthine cord. Lately, reports had it, the seam had been growing thicker, thicker . . . had widened to five feet, to a dozen, to fifteen . . . Rumor even mentioned thirty feet and sixty, but that was stuff for laughter. Who would believe that Mackay and Fair were nearing a lump of solid treasure 700 feet long, five hundred wide, ninety to three hundred high, so rich that it would demonetize silver, destroy men's minds and give shape to the national politics of the next sixty or a hundred years? Yet on reports one-tenth as bizarre as what turned out to be truth, Con Virginia's shares were already rising.

Meanwhile, at the surface, three factions were squaring off for battle. These were Jones and Hayward in historic opposition to the long-reigning Sharon and his Ralston-Bank of California crowd; Fair, Mackay, Flood and O'Brien

blocking out their own illimitable bonanza and staving off
the Sharon-Ralston ring who wanted it; and in solitary
grandeur an indomitable Jew, Sutro, who was relentlessly
driving a four-mile horizontal tunnel from the hillsides
above Carson River that would tap all the mines at depth,
drain them of heat and water, and—as the magnates feared
—drain them also of the profits of market-jugglery and
monopoly.

Elsewhere the year was closing on national depression,
the aftermath of Civil War and its financial excesses. But
San Francisco was brilliant, light-hearted. Its streets were
gay with carriages and well-gowned women. Its helter-
skelter buildings of the Gold Rush days were being replaced
in spacious stone. Ralston of the Bank Ring was clearing
ground for his Palace Hotel, the like of which no city had
ever seen. The railroad quartet who had completed the overland Central Pacific four years before were eyeing Nob Hill
for their ostentatious palaces. Cafes, theaters, dicing halls
were crowded. It was as if all felt the orchestral tremor of the
speculative dance that was about to begin, making feeble any
such dance that had gone before.

San Francisco, after twenty-four years of Mother Lode
fervor and fourteen of Comstock, was about to play the boards
for the Big Bonanza.

In this kinetic hour, the headachy man who had shaken
the hand that held half of Crown Point's millions must have
felt that Panamint was very far away.

# 6

## SHOO FLY

Snow had begun to fall when Raines and party had moved down Surprise Canyon two weeks before with their ore samples. It was forerunner to one of the heaviest blizzards known to the intermountain country in ten years. On December 12th the mercury had already dipped below zero. Sierra passes were snowbound. Out on the sagebrush plateau, high winds tossed up sand-pillars a thousand feet tall that bored over the landscape like augers and churned between world and sky like giant cranks. Snugly arrived at Los Angeles, the promoters were out of that hard, dry cold and biting gale. The southland village became aware of the extent of the tempest aloft when Nadeau's Cerro Gordo freight teams failed to get down to the coastal plains on schedule.

The residents of Panamint, who were as far above the desert dust-whirl as Los Angeles was below it, nevertheless dug themselves in for an alpine winter. From the inexhaustible boulder supply they raised up cabins, hewing rafters and shingles out of the scraggy timber. Some shelters were built paling-fashion out of vertical logs. Honest men and rogues dwelt in cheerful comradeship. The greasewood telegraph continued to chatter of great things ahead, and by March, despite rigors on the road, there were seventy people.

Newcomers were hailed with gusto. "Look at them ledges,

63

stranger. All we need is capital. Big operator from San Francisco is tending to that part of it. Roll yourself together some stones for a house and watch this camp grow."

The returning Kennedy was one of the first to get through the passes. On his way from Bakersfield in January he stopped at Havilah and gave out an interview concerning the new silverado which Havilah's *Miner* graciously gave circulation to:

"No mining expert has yet visited this mining district (and there have been many sent by mining speculators) who has not expressed wonder and astonishment at the immense amount of silver ore in sight; and several, who have mined for years in the State of Nevada, have expressed the opinion that Panamint is certain in a few years to eclipse, in the amount of silver yield per annum, the famous Comstock lode of that State."

But days passed by and the capital promised by Raines did not appear. Neither did that worthy. His resources were strained to achieve a meal and a bed. Vanderlief seems to have failed him this time. Panamint, to the man adorning a San Francisco curbstone, must have taken on a kinship with the mirages. Something once brightly glimpsed, then lost, like the Gunsight mine or the Goller or the Breyfogle. What a night he had made of it . . . what a morning waking . . .

*Shoo Fly. Shoo Fly Landing.*

A fellow operator sitting in the bright sunlight beside Raines abandoned his *Daily Alta California* and went for a stroll toward a pair of swinging doors. Odors of the free lunch were appetizing. Raines would have liked to follow, but his credit there was at the moment in abeyance. As for cash . . . Well, a man can at least keep posted as to the news. Raines reached for the many-columned *Alta* and perused it.

*Rancho Santa Monica. Acres. Sea view. Shoo Fly Landing.*

Raines brushed this persistent shoo-fly aside and turned to other columns. Comstock's production to date was estimated at two hundred millions and the splendid holes had already gulped back the major portion of it in assessments. The new Southern Pacific was pushing south from San Francisco Bay and was also fiercely contesting the Texas & Pacific for a federal subsidy for its proposed tracks from El Paso westward. J. P. Jones, that independent genius, was fathering large plans before Congress for reclamation of the southwestern desert. He had surveyors in the field studying the practicability of flooding all the below-sea-level country, creating from the overflow of the Colorado River a chain of navigable lakes whose presence would enhance fertility, ameliorate temperatures, suppress hot parching winds and sandstorms, augment rainfall and create a region rivaling the valley of the Nile. Now if a dam could be inserted to plug the riotous Colorado, considered this far-seeing Nevada solon fifty years ahead of his time. . . .

*Baker & Beale. F. P. F. Temple. Shoo Fly Landing. Los Angeles. Independence.*

Raines continued his inspection of the news of the day, spurning this chit-chat of rails and real estate from the lower end of the state.

Silver's demonetization by Congress was being scathingly denounced by William M. Stewart. Silver's demonetization was to be further denounced in a forthcoming speech by John P. Jones. Jones was buying mines, ranches, newspapers. Jones was believed to have ambitions to own a continental railroad of his own. Jones was in Wall Street. Jones was in horses. Jones was in love. Jones, Jones, Jones.

*Shoo Fly Landing. A railroad through Los Angeles from Shoo Fly Landing to Independence.*

As if the shoo-fly had become a gadfly, Raines started. A panorama had opened out. It was a panorama of breathtaking breadth and splendor, with plump, bald, brownbearded John P. Jones hovering over it. A Jones afloat on silver wings, his toga streaming out behind, his ruddy countenance suffused with benevolence and his arms emptying wide-mouthed money bags.

The pattern of it all grew plainer. Col. Robert S. Baker was a San Franciscan with a fortune made out of supplying the mines, now a heavy investor in real estate. Baker was associated with E. F. Beale, a former surveyor-general who had come into possession of broad sheep and cattle ranches. Lincoln himself had observed, after seeing how deeds and acres flowed to this practical civil engineer, that Beale was certainly "monarch of all he surveyed." Baker and Beale had lately purchased from its Spanish proprietors the 38,000-acre Rancho Santa Monica y San Vicente overhanging the sea near Los Angeles. Now Temple of Temple & Workman, those bankers whose establishment adjoined the Clarendon Hotel in the southern pueblo, was associating himself with Baker and Beale. The three were projecting, on paper at least, a narrow gauge railroad from Shoo Fly Landing on the Santa Monica Rancho up into the deserts and on to Independence.

Independence was the county seat of Inyo—the county that contained Cerro Gordo, Death Valley, Panamint.

The vista was growing stupendous. It took account of half the geography and politics of the West.

J. P. Jones, Raines knew, despised the Central Pacific crowd as a normal commoner hated greed and arrogance in high places. The Stanford-Crocker-Huntington combination feared and disliked Jones because the Senator sought to tax their federal land grants, and because he was known to favor competition for their giant railroad projects. From Shoo Fly Landing to Independence was 225 miles. If carried on to Ogden, five hundred miles farther, Temple's proposed Shoo Fly line would tap the Union Pacific, provide the Coast with a new overland artery, and divert much travel from the Central Pacific straight down through the southwest to Los Angeles. Moreover, the new artery would pass close to Panamint. J. P. Jones, with his numerous ambitions, could hardly fail to become re-interested in Panamint.

The pattern was complete. Raines was up and on his way to interview Colonel Baker. Other operators along the sidewalk who clutched for his *Daily Alta* may well have searched its columns to discover what had caused his fast departure.

Colonel Baker had, in his Santa Monica Rancho, a dream terminus but as yet no railroad. The Senator off in Washington had the dream of a railroad but no terminus. Raines had, in Panamint, a half-way station that promised a world of freight. Clearly the Senator, whose money-bags were the most substantial item in this figment, must be made to perceive that Panamint was the logical link joining Shoo Fly, Los Angeles, Independence and the unconquered deserts clear to Ogden in a grand imperial development.

Baker listened to the impecunious schemer, had to admit the possibilities. There was a rebolstering of Raines' finances, followed this time by no dionysiac celebration, and the spokesman for Panamint found himself bumping east on

Stanford & Co.'s Central Pacific, bound for Washington.

The approachable Jones would listen to anybody, even a Raines, and to any proposition, even twice. A few weeks later the *Index* at Santa Barbara on the Pacific shoreline was telling its readers:

"We are, by good authority, informed that a brother of Senator Jones, United States Senator from the State of Nevada, lately visited the Panamint district, 90 miles southeast from Cerro Gordo, and found the mineral discoveries there all and more than represented in character, width of veins, and in richness. He purchased five claims for $113,000, and has gone up to San Francisco to make preparations for the thorough working of his mines. We further learn that there are already over 100 persons in the Panamint camp, and that many more are going in from Los Angeles, San Bernardino, Bakersfield, Havilah, Cerro Gordo and Independence. The indications are that Panamint is the richest mineral strike ever made in quartz on the Pacific Coast."

# 7

# THE COMING OF THE BAD MEN

On September 17th, 1871, while the jailer was going about his evening lockup duties in the Nevada State Prison at Carson City, twelve of his convicts suddenly charged the guards, broke into the arsenal, seized guns, liberated seventeen other train and stage robbers, slew three men, and started down the California-Nevada border on a career of burning, shooting and pillage. Most of them were wiped out in bloody set-tos or escorted back to jail again.

From Carson City to the valleys under Telescope Peak is 250 miles as the hawk might fly, and as much farther for rabbits, human and otherwise, as doubling and twisting might dictate. But the rabbits in this case were desperate and the goal was safety. Rumor had it that some of the escaped men were making for the folds of the Panamints.

Cerro Gordo was a camp that had no name whatever for civic culture or orderly existence. It was given wholeheartedly to that singular appetizer, a man for breakfast. Yet even in Cerro Gordo a man could become too indigestible. Whereupon his fellow-residents would take him to a place of large view, show him the beetling Sierras forty miles away on the one hand and the sand-ringed Panamints forty miles away on the other, and bid him make his choice—but choose. Occasionally the citizens so propelled had already been ejected from Havilah, Kernville and the other camps in the Sierras, so they took the route eastward as the best alternative.

It had come to be recognized, therefore, that there were sundry inhabitants in the Panamints who had quite hurriedly left their spare shirts and all shaving equipment behind at Cerro Gordo.

Off in a distant northern arm of Death Valley there were a couple of camps known as Lida and Gold Mountain. Occasionally Lida and Gold Mountain ran some of their residents out for cause. The discards had likely enough been previously run out of camps east, north and west. The only direction left to them was south. If the month wasn't June, July, August or September they had a good chance to live. High above the relentless valley which lay on that southward track the Panamint Mountains offered wood, water, and a small amount of game.

So the understanding in Lida, Gold Mountain, Benton, Columbus and elsewhere was that mail unclaimed by certain hastily departed townsmen might reach them—this was high humor—if readdressed Care of the Panamints. A little spring on the edge of Panamint Valley, bubbling under some thorny mesquite bushes, was known as Post Office Spring. Once in a blue moon some passing friends left tobacco and supplies at this desert well for the outcasts above.

Occasionally there came sanctuary-seekers from greater distances. One of the prime industries of the sagelands, for fifteen years past and twenty years to come, was that of touching up the stagecoaches. The stages regularly carried Wells Fargo's express boxes, resting under the feet of drivers whose hands were reasonably busy with lines running to several horses, and who had no time or inclination to shoot things out with men who suddenly bobbed up out of the shrubbery with a "drop." But an industry second only to stage-robbing was stage-robber chasing. This was because Wells Fargo had a standing offer of $300 for any of its highwaymen, whole or in pieces, together with a quarter of any rescued treasure.

The result of these many reward-postings was an expanding circle of settlements to which the men with guilty consciences gave wide berth, and it was deemed probable that the Panamints afforded to some a grateful hideaway.

Austin and Eureka, White Pine and Pioche, were other districts that had all they wanted of a certain class of citizens and sometimes urged these to go look at the view. Pioche in particular was practically seeded and fertilized on bloodshed. There was a preliminary affair concerning a mine known as the Washington & Creole, an outcropping of hillside ore worth $300 a ton. A pair who had a claim on the slopes above decided to jump the Washington & Creole. For this purpose they built a fort in the night and sent for some of the toughest bullies in the White Pine district to come and help man it. To Raymond and Ely, who owned the hijacked Washington & Creole, appeared four young men who said: "Give us permission to work that ground for thirty days and we will drive those fellows off." Fortified with a written promise and liberal whisky, the quartet thereupon launched a drive on the hill, and in the flurry of shots killed a bystander and took possession.

Fifteen thousand dollars each was the profit made by the group on their ensuing thirty-day lease, though Pioche was in continual uproar and scores of fighters were imported at wages of $20 a day to keep the rival powers in balance. Of the four who seized the fort, one was shot and killed the following year in Eureka, one stabbed a man and left camp, one—Morgan Courtney—became a gambler and ruling czar of the toughs; and the fourth, Mike Casey, set off a long train of fuses in the following Pioche manner:

Casey owed a friend $100. When he had lugged $15,000 to the bank from his month's work on Washington & Creole, the friend dunned him for the hundred and accumulated interest. Casey conceded the principal but refused the interest. Gunplay followed. The friend went down, leaving a will that was something like a time-bomb, for it pledged $5,000 to whomso would up and kill Casey. The armed defense of Casey by his friends, and the attacks on his defenders, would fill a book and have no place in this narrative, save that they caused numerous men from time to time to leave Pioche with more or less abruptness and head for the Panamints.

How many such reprehensibles dwelt in the tortured, upheaved range nobody knew, because nobody looked. Senator Bill Stewart, who later came to know some of them well, wrote them all down as "bad fellows, outcasts of society, who obeyed no laws, not even their own, for they were not organized into a 'gang,' but practiced their profession in an entirely independent manner." Notwithstanding, they made a go of neighborliness all through that first winter which proved quite a contrast to later times. C. E. Krause, one of the reputable pioneers of Panamint, wrote to his family who kept the stage station at Little Lake:

"I think we have in this camp the most intelligent and liberally inclined miners that perhaps ever got together."

Winter lifted. The greening uplands just beneath the receding snow line were as a signal flag to the desert's far-scattered nomads. A few turned their burros that way. They found the hamlet of Panamint still a huddle of hovels—a strike, perhaps, but not yet a boom. Off in the north played that lurid light, the white and blinding brilliance of the Comstock. Dimmed by such sky-filling radiance, Panamint village was scarcely a star on the opposite horizon.

Then came the news about Panamint, this time not tossed from screw-bean mesquite to rabbit bush but rather from newspaper to newspaper the length of the Coast, that J. P. Jones was buying.

# 8

# THE COMING OF THE COMET

WESTWARD from Pioche rode Dave Neagle, his packmules laden and his spirits light. Neagle years before had run away from Santa Clara College to see life as it was lived in the Nevada silver camps, and boy and man had seen a good deal of it—by now having reached his twenty-seventh birthday.

That he had lived so long, when better men had died younger, was due to natural quickness on the dodge and cultivated speed on the draw. These were qualities much esteemed in White Pine and Pioche. Lately his quick hand had enabled him to drill an adversary through the cheeks but misadventure had followed. The sheriff had collared them both and locked them up together in the same cell. The purpose behind this, Neagle suspected, was less to purify the civic scene than to tickle the Pioche sense of humor. But any gunman would rather be felled outright than served up for public laughter; besides which, each hour in that cell had consisted of thirty-six hundred seconds of apprehension. For Jim Leavy, with that hole through both cheeks, simply wouldn't be friends. . . .

Once out of that, the sheriff having had his peculiar fun, and Neagle was moving at a good traveling pace toward new scenes and prospects.

The long trail wound through vales that were velvet with young sage and past slopes that were aflame with the scarlet torches of the hundred-stemmed ocotillo. The desert wren scuttled ahead of him to her eggs beneath the harsh-barbed cholla cactus. The giant suhuaro, which takes a century to grow to the height of a man and then stands centuries more, confronting and tossing the windstorms like a wrestler, pointed him onward with its angular arms.

Through the country of the sullen and unaccountable Tempiute Indians. Past the Belted Range. There were always more miles ahead. Wells were few: Summit Springs, thirty-five miles beyond Hiko; then twenty-two to Quartz Springs; then thirty-five to Indian Springs; then to King's Ranch, fifty. Dave Neagle watched the night-blooming cereus, crept upward from the Arizona country, for his calendar. And presently with a summer midnight those spicy waxed flowers unfolded. In a single hour, as Neagle looked on, they reached their prime. Were they symbolic in their showy brevity of the Panamint that lay ahead? This Neagle could not know, but he did know that it was the twenty-fourth of June, 1874, and over-late for man or beast to be where he was riding.

The greasewood slopes, with green mesquite along their bottoms, gave way to starkness. This was the Amargosa region. By day its biting pools lay sun-hot. The song had subsided on Neagle's lips. These surrounding hills might be shot full of treasure, but Neagle was no prospector. Thirty-eight miles lay between King's Ranch, last known spring, and the place where he would camp tonight. Presently his horse's hoofs were ringing down the red stone stairs of a descending corridor. The walls were closing in and lifting straight up. Eight twisting miles through and adown their

76 *Silver Stampede*

colored bandings brought the rider and his pack train to the bottom. Here the shade was cool and intense.

Neagle was now on the below-sea edge of that "Valley of Bones" whose oven-dried air at midsummer brought insanity and death to inexperienced traversers. Opposite sprang the Panamint Mountains, reaching for non-existent clouds.

Neagle was desert-wise, and he held his animals in the shade of the red corridor until another nightfall, watering them at a basin scooped from the gravelly floor. Then he struck out across the wide, white sink, which was currently reputed to be devoid of water or vegetation and never darkened by shadow of passing bird.

As for nights past, a great comet was burning overhead, rushing with spectacular brilliance through the sky. The celestial searchlight flung its beam from the northeast horizon nearly to the pole star. Fifty-three million miles away, just within Venus' orbit, it lighted him onward over sulphurous wastes that Neagle had no desire to know at noonday.

Dawn saw the traveler ascending the farther rim. New night found him just over that western summit, refreshing at a big spring that was fed by patches of high, remote snow. Another dusk found him at that canyon spring sheltered by its grateful wild rose where Henderson and George had camped in the decade before.

Two days later, by a long drop down into Panamint Valley's desert and the steep climb up Surprise Canyon, and Neagle of Pioche rode into the new town of Panamint.

The settlement he found there was a string of rude dwellings along a wide, vacant street in which no mule team, bullock team or stagecoach had yet made swing. There were

*Drawn by E. A. Burbank.*

Panamint Valley and Maturango Peak in the Argus Range, from the entrance to Surprise Canyon.

"Opposite sprang the Panamint Mountains."

no stores, stables or accommodation for man or beast except, for the latter, wild grass on the hillsides and for the former a big tent known as the Hotel de Bum. No charge of any kind was levied for entertainment at this tent and no one seemed quite certain as to who was the proprietor, but it had a cook of sorts, rough grub and a hearty welcome and rumor said the invisible host was a well-to-do senator at Washington.

Neagle had that on board his mules which made his coming a historical moment and pronounced the town officially opened. He set up two barrels and a board, with a couple of glass tumblers for elegance, and a rough sign indicating that the Oriental Saloon was ready for business.

Close behind Neagle, similarly equipped but moving under the light of the comet from the opposite direction, came genial Ned Reddy, who was declared to be entitled to six notches in his well-polished gun handle but who deprecated the distinction. Once in Aurora a townwide cleanup had left several men dangling on a vigilante gibbet. If Ned Reddy and his brother Pat reached the next camp, Mount Montgomery, a little out of breath, it could be attributed to the high altitude. While Pat was getting an arm shot off at Virginia City, Ned had drifted to Cerro Gordo where he opened a gambling house. In '71 he had bagged one Tom Dunn at Cerro Gordo but the inquest had adjudged it self-defense. In '73, it was affirmed, he had shot Bulger Rains at Columbus just to see him squirm. But he had been slightly confuscated at the moment and, anyway, the Rains in question was of little value to the community. That made the total notches two, and six was plain exaggeration. Ned Reddy, who was handy with a guitar, planned to make his Independent Saloon in Panamint a jovial resort and didn't want a reputation that might scare away customers.

Behind both, his way illumined by the same comet, came James Bruce, a short, stocky, dark-complexioned Kentuckian. Bruce thought that a camp of so much promise as Panamint might have some use for an undertaker who was also, between jobs, a skillful faro dealer. He made no point of the fact that he was also unbelievably quick with his six-shooter, the wide muzzle and well-oiled mechanism of which had already taken off one bruiser and would presently take more.

In Bruce's tracks from the east came Billy Killingly, soon a long way from Pioche but still moving rapidly; Billy had got out of town and a considerable way on the road to Panamint before chancing to discover that he had a number of horses in his string belonging to other people. He hoped that the owners had been as slow to discover the error as himself.

Up the canyon from the west came also small Tom Kirby, who had been herding with the Morgan Courtney gang at Fish Springs a hundred and fifty miles northward. Morgan Courtney, Kirby's old chief, following the Washington & Creole battle at Pioche had become a professional strong-arm whom mine owners over a wide territory hired to come and clean up labor and claim-jumping conditions. He maintained his own wrecking crew of toughs and it was said that his unwrapped rifles and revolvers, bearing his name neatly engraved, when shipped into a town by Wells Fargo were usually enough to quell any disorder, whether or not their owner followed in person. But sudden "lead poisoning" had lately overtaken Morgan Courtney at Pioche, and Kirby, left without his chief, had heard the firm suggestion that he start riding. Kirby had ridden in consequence for Panamint.

From this direction and that came others, laying their own courses and blazing their own trails; Telescope Peak was a landmark visible for seven-score miles, and the brilliance of the comet shone for all.

The precious pair who were to put the ultimate seal on Panamint's reputation were a little slower in arriving, being detained by business on the road. But finally they too were present.

John Small and John McDonald were New Yorkers who some years before had faced west, motivated by a desire to pull the frontier apart and see what made it tick. They discovered that primarily what made it tick was Wells Fargo's stages, which bumped across the landscape with regularity and carried nearly all the West's mails and bullion. At Denver they began to hear about Panamint, at Salt Lake City they heard about it still more, and at Battle Mountain it became borne in upon them that Panamint was what they had been heading for all these months without knowing it.

To get to Panamint from Battle Mountain you transfer to the stage, wind southward up Reese River for ninety miles, get down at Austin at an elevation of 6,000 feet, look over the local bank and the daily bullion shipments of the Manhattan Silver Mining Company, steal a couple of horses, and continue south over no roads at all until you have passed the Toiyabe Range, crossed Cedar Mountains, and skirted the forlorn spots where sharper eyes will someday detect Tonopah and Goldfield. Then you have your choice between crossing Death Valley and the Panamints, or swinging right and entering Death Valley's companion, Panamint Valley, from its northern end.

Small & McDonald reconnoitered Austin for a day or two, laid in supplies for the forthcoming journey, realized that they would see no more stagecoaches for many a day, tied their horses in the sagebrush, relieved the northbound vehicle of its bullion cargo, helped themselves to whatever else was in Wells Fargo's coffer, and waved their late conveyance on its way.

Their further progress was speeded by word, tossed from mesquite thicket to creosote bush, that Panamint was soon to have a bank.

Small & McDonald were specialists in that line. They were in Panamint town before the bank got there. They filled in the interval by prospecting, and to their pleased astonishment uncovered, a few miles north of camp, one of the handsomest silver ledges in the whole vicinity.

Though short of actual cash, Small & McDonald perceived that the world could be very fair of aspect. They now had three assets: a mine which they had no intention of digging, a prospective bank in which others would supply the deposits, and sooner or later a line of stagecoaches that would have to come and go through a weirdly steep and isolated gulch— coming, presumably, with travelers bearing fat wallets, and chests full of coin, and departing with Wells Fargo satchels full of bullion ingots.

Never had the pair imagined such a perfect all-round setting for the practice of their peculiar art. Let others blast at the cliffs, shovel ore, mill it and transport it and sweat about it generally in all the lather they pleased. With a merchantable claim, practically a private bank of their own and an ultimate line of stagecoaches that would virtually have to lean back on their hind boots while ascending that winding, narrow gorge—with all these things in prospect, Small & McDonald found the outlook simply admirable.

# 9

## THE ROAD UP THE CANYON

PANAMINT was now approaching the status of a minor excitement, but getting there was still considerable of an expedition. The new camp was at the apex of a 160-mile triangle whose other corners were San Bernardino and Bakersfield. Each of the latter towns was a coaching terminus close to the base of the Mohave tableland. But the Panamint Mountains were far across that tableland and roads were non-existent.

Yet those who beheld in every new strike a potential Virginia City were already on their way.

Those from San Francisco, via Bakersfield, had the benefit of the stage line over the Sierras as far as the Owens Valley. Those from Los Angeles and San Bernardino had a lengthy crossing, once the plateau was reached, over dry lakebeds and around chains of ocher, mauve, jet and magenta mountains.

From Carson and Virginia ran a stage line of sorts down the east footings of the Sierras to Owens Lake. Its passengers joined the Panamint migration at the lower end of that peculiar dead sea which so lent itself to strange fountain-like antics during earthquakes, and in whose saline depths more than one passing teamster, illuminated by a cold desert moon, insisted that he saw scaly-backed monsters rolling like dolphins, slithering like serpents and frightening his mules like hell.

All routes for Panamint, after leaving the established stage tracks, led out into a high, trackless, deep-gouged world fit only for the sturdiest saddle-horse and the shambling packmule.

But the hour for roads to Panamint was nearing. R. C. Jacobs in March had led a crew down from Surprise Valley and attacked the roughest spots with picks and shovels—proceeding then to Los Angeles to fan the campaign which Raines had started.

The solid character of the prospector and the vigor of his assertions whipped up new interest at the pueblo in the orange groves. Jacobs pleaded that the miners were doing all they could to promote ingress and egress, but that they were ore-rich and cash-poor at the moment and needed help. He reminded anew of the immense commerce that should flow to Los Angeles, but pointed out that the Southern Pacific Railroad, by building southward up the interior Valley of California, would be able to claim this traffic for San Francisco Bay.

For $40 a mile, Jacobs believed, a serviceable road could be carved between Surprise Canyon and the Cerro Gordo–Los Angeles trail. Echoing Jacobs, the newspapers urged that Panamint's broken ledges already disclosed riches far beyond those first revealed along the Comstock. Southern Californians harked, and again contributed.

Meanwhile a toll road up Surprise Canyon had been undertaken by Bart McGee and four fellow Panaminters. The promoters had nothing to pool but their brawn and pluck, so they pooled those, and the spiraling gulch became dinful with the pushing apart of walls and the beveling off of precipices. As the five men toiled, Coggia's Comet continued its nightly fireworks through the sky. Its tail by now reached far beyond the pole, its expanded form filling all the northern heavens like a luminous cloud or an immense new Milky Way. Its presence was awe-compelling over a land already sufficiently mysterious. To the superstitious it presaged disaster in countless forms. Cloudbursts of excessive fury, for example.

"Nonsense," said Bart McGee, and drove his companion road-builders forward under the heat. The last of the rocks were riven apart with a walloping powder-clap that was saved for the Fourth of July, and Pat Reddy of Independence, lawyer-brother of the guitar-toting Ned, made the first wheeled ascent of the spillway in a buggy.

An advocate attuned to robust times was Pat Reddy, permanently prepared to battle or defend anything that wore hair or pulled hair trigger. After quitting Aurora in that exodus of unwanted sports and becoming embroiled with a faster man at Virginia City, he had retired to Independence at the foot of the Mount Whitney country, acquired a few law books and made himself an attorney. Though the weapon waved by Jack Mannix had cost him one arm, with the fist attached to the remaining member he could pound the bar of justice so vehemently that he was credited, first and last, with freeing more than one hundred killers in court-rooms east of the Sierras. Utterly brave, cocksure and unscrupulous, Pat Reddy was the criminal's unfailing refuge and the prosecutor's despair; of him it might well have been written, as was said of a contemporary barrister, the strong-willed William M. Stewart of Virginia City: "Once enlisted as counsel in a case, he made the cause of the clients his own. He saw no foundation of justice in any claim of an opponent and left no stone unturned to achieve success. His own determination to win at any cost, and the belief that he could match his adversary with any weapons . . . exposed his course to sharp, if not merited, criticism; but he defied his critics." Between passages at law, Pat Reddy dabbled in mining properties.

C. A. Stetefeldt, an engineer of standing, also went up Surprise Canyon in a buckboard and made an emotional report about it all to the United States Commissioner of Mining Statistics:

"In ascending this remarkable canyon one is surprised indeed; such steep, bold and barren mountains, intersected by deep gulches; such a variety of rocks; such grand traces of the work of the unfettered elements! I can only compare it with a chart of the moon, and conceive that such must have been the aspect of the whole of our earth in its earliest state."

He added, and his report fanned the spark: "It is rarely the good fortune of a mining engineer to form so favorable a judgment of an entirely undeveloped district as I feel justified in expressing in regard to the mines of the Panamint District."

Late in that month of July, 1874, overland railroad passengers near Winnemucca, three hundred miles away, were watching with interest as their train left Mirage Station and entered the illusion of a sparkling lake.

As the tracks behind them were seemingly engulfed, and travelers crowed with delight at the pretty phenomenon, the lake suddenly blotted itself from sight. The sky had grown dark. Black clouds smote the summits of the nearby hills. Before the engineer could crowd on steam, his train had been gulped by no mirage. The downrush of waters swallowed his fire and flung his locomotive from the track, fortunately without loss of life.

This derailing of a train by actual liquid in the middle of

a mirage was but an incident in an unprecedented series of cloudbursts which just then fell upon the Great Basin. Before evening of the same day another train, near the Utah line, was hurled from its track with death to five passengers.

In this same hour a frightful target was made of Eureka in the center of the sagebrush state. From early morn until noon of July 24th, rain had been falling with violence. Then the sun came out, to dim again as a new rainbag burst against the head of the valley.

In ten minutes the flood was feet deep, and human chains were formed to drag inhabitants of the canyon town to safety. In twenty minutes the little community had lost half its buildings and fifteen lives, including women, children, Chinamen, and the editor of the *Cupel*, who was swept away with his types. The colossal storm then swung far south. In Arizona the Hassayampa and Verde rivers went on rampage; Chino Valley near Prescott was engulfed and the Twenty-third Infantry, encamped to fight Indians, saw its tents and baggage go down stream under such a deluge that it might have been pardoned for thinking it had joined the navy.

The cyclonic "waterspouts," venting their wrath hundreds of miles apart, coincided with the earth's passage through the tail of the comet. The planetary visitor was likewise being held accountable for many other contemporary griefs—the kidnaping of a little boy, Charley Ross, from his home in Pennsylvania on the first day of July; the burning of a part of Chicago on the fourteenth; the scandalizing suit of a choir singer's husband against the Rev. Henry Ward Beecher, a conspicuous divine in Brooklyn; and the divorce suit launched at Brigham Young by his nineteenth wife.

Without attempting to explain these other woes, hard heads rejected the comet as the cause of the Great Basin storms and ascribed those affairs to the tremendous mountain snows of the previous winter, acted upon since by months of water-lifting heat. While argument raged, similar summer tempests plunged Hamilton and Pioche under a foot of water and all but washed away Elko and Austin. In the White Mountains above Benton, drawing nearer to Panamint, a rainspout fell upon a mule team and swept it for yards along the road, killing four animals while the driver escaped with his life. Below Owens Lake a thousand cords of wood and tons of charcoal intended for Cerro Gordo were washed away and scattered over the plain.

Billy Killingly, who was following Dave Neagle's general route from Pioche, elected to make his way into Death Valley around the south end of the Funeral Mountains. Billy found the way broad and smooth and a late July evening brought him to a wide place of such inviting levelness that he concluded to throw down his blankets in the middle of it and spend the night.

Unknown to the sleeping man, stars and the comet went stealthily dark. Clouds burst against Black Mountains and Funerals. Their waters sprang down both sides of Furnace Creek Wash, converged and charged for his bed. Billy leaped to his feet when he heard the oncoming roar, abandoned all gear, and made with might and main for higher ground.

Before he had traveled a hundred yards the water was at his ankles. Before he traveled five hundred it was up to his hips. Billy flung himself to the stones of the steep slopes after a struggle. There he waited out the night, thankful to be alive. Death by thirst in Death Valley he was ready to chance, but extinction by drowning had not been in his calculations. An hour or two after the sun rose, no trace of the night's flood remained near the traveler's late sleeping quarters—and no trace of his belongings. Horseless, shoeless and breakfast, the man made his way to the mouth of the wash, where by good fortune he found Andy Laswell and Cal Mowry, also of Pioche, wrathfully gathering the scattered belongings of their camp.

In such a carnival of the elements Panamint was not neglected. Those clouds that opened over Furnace Creek Wash had enough left in their embrace to drench the crags of the Panamint Mountains. New-born brooklets leaped down the three-sided bowl and met above the tiny town. The usually dry storm channel grew suddenly bank-full. Stockades, thatches and tents went voyaging.

With joy in the destruction, the waters rushed on and made their leap for Panamint Valley thousands of feet below. Anything that remained living in the wagonway did so by grace of scurrying up the sides. With a roar of tidal grandeur the flood bore down on screaming horses and lashing jacks.

Just twenty-two days after that clap of gunpowder had announced the new road open, the labor of months had vanished in fifteen minutes. Far out into Panamint Valley Desert the material was borne, to be dumped halfway toward the Argus Mountains.

The sun came out. The rocks of Surprise Canyon sent up lively steam. Off in cosmos the comet gathered its gaseous train and rushed splendidly on.

Bart McGee and his partners, with a curse for desert cloudbursts, set doggedly to grading their road again.

# 10

# THE COMING OF THE RAINBOW–HUNTERS

THE rush was on now. Fortune-hunters were pressing by horse, mule, buckboard and afoot. There are half a dozen classes of mining men known to the western country —the solitary prospector, a roving hermit whose joy is in the quest and who, locating his strike, promptly sells out, packs up and moves on; the gregarious chaser after every rainbow, who hurries to be in on each new stampede, ever dreaming of staking out a rich location next to the discovery claim; the placer miner who plies pan, rocker or long tom along some creek, working for himself alone—a breed that pretty well died out with the first years along Mother Lode and Comstock, though revived again by the distresses of the Nineteen Thirties; the promoter bristling with hokum, plans and prospectuses; the absentee owner; the simple laborer, who mucks or blasts for the latter for his $4 a day; and the medley of merchants, restaurant keepers, resort keepers, gamblers and camp-followers who throng to jostle and breathe the optimistic air of each new settlement.

Panamint was different only in that the wandering prospector had been preceded by other nomads who, coming upon surface color, had sat tight with their uneasy consciences and wondered what to do. But the word was now out, and from a dozen directions the seekers for the pot of gold—silver would do quite as well—came hurrying at the beck of the Panamints. They pushed past even as Bart McGee and his fellow-townsmen toiled.

Eternally on the move, they had raced each other before for Gold Hill, Silver City, Virginia City, Esmeralda, Reese River and White Pine. They would jostle again for Bodie, Tombstone, Deadwood, the Cœur d'Alenes. Mormon Bar and Whisky Flat, Fraser River and Powder River had known them; they would yet scale frosty Chilkoot Pass together; they would scuffle and sweat in Goldfield and Rhyolite. The deserts were already crisscrossed with their trails, monumented with the rusting boilers and gaunt mill chimneys of their transient hopes, strewn with their bleaching cabins and the bottles and cans of their bivouacs.

Would Telescope Peak be another Treasure Hill? Veterans of the rush to the White Pine district asked each other hopefully, cynically, humorously, according to their various temperaments, as they pressed on. In just such a place, some remembered, an Indian had brought a chunk of rock to prospector Al Leathers and led him to the rich chlorides of the Hidden Treasure mine. The result had been Treasure City, and the uncovering of that pocket known as the Eberhardt, seventy feet long and forty wide, from which $3,200,000 had been lifted. Thirty-two hundred tons at $1,000 a ton, and not a pound of it below a twenty-eight foot bottom. No wonder the population of Treasure City had clung to its huts and caves nine thousand feet above the sea, while the thermometer sagged and smallpox raged. But all that was five years ago, and these men could do now with another bonanza.

Sam Tait, at his station far down Surprise Canyon, cheerfully filled up man or beast and collected the road toll from

each—$2 for a wagon, four bits for a horseman, two bits
for a packhorse or jack.

Up the canyon came Bill Raymond on a strong fine mule
and a good saddle; Bill Raymond of Pioche, no less. Fate
had grabbed Bill Raymond by the ears and bounced him
up and down so often that he was always more or less ready
for her next grab and jounce. But just now he was up. Over
near the eastern border of Nevada the mighty Raymond &
Ely mine was vomiting forth its $17,000,000 worth of riches
and Pioche, the town that clustered around its shaft, was still
talking about how Bill Raymond once gave the Mormons a
silver watch as down payment for that lordly claim. The rest
of the $35,000 needed to pay for the mine had been taken
out of the ground itself in sixty days and John Ely had sold
his share for $350,000 while Raymond stayed in. All from a
sixty-dollar watch, reflected Bill Raymond, that always did
run a little slow. Well, it hadn't made him late in Pioche.
Would the chronometer, long since redeemed, get him on
time to Panamint?

Up the canyon came a character known as "Old Tex"
Shore, who had lived such a variety of lives that a newspaper
once offered him $1,000 to write up some of them—Old
Tex, after thinking about it, refusing; there wasn't one that
he dared to commit to paper. After an odyssey that had
ranged from Coloma to Bannack, with calls on sirenic Virginia City, Boise, Silverado and Treasure Hill, Old Tex
reached the mouth of Surprise Canyon with just one valued
possession in the world—a monstrous, breech-loading, snap-
action Parker shotgun.

"I kin hit," said Old Tex at Sam Tait's place, "a sagehen at a distance of a hundr'd yards. But I gotter be loaded."

Someone "loaded" Old Tex with what his personal muzzle

needed, and then went out looking for a target. Long before one offered itself, the effect of Tex's private charge had worn off, and he demanded more internal explosives.

"I kin hit," he proclaimed, patting his lusty fowling-piece, "any sagehen with Old Bess here at a hundr'd 'n' twenty-five yards."

Again the proper charge was rammed home to Old Tex's vitals, and his audience waited for a bird. Before one appeared, Tex was really ready to assert how Old Bess could shoot.

"I kin hit," he promised, "any object that runs 'r flies at a distance of a hundr'd 'n' fifty yards—not an inch less."

This time a gay-crested roadrunner was spied down the slope at just about the distance Old Tex craved. The arrival was busily turning over stones and cakes of mud for crickets and millipedes.

One more charge for Tex, a mighty clicking of Old Bess's hammers, and the long weapon was up to its owner's eye and shoulder. For a moment he held it there, and all stood waiting. Then he lowered the fowling piece.

"Shucks," he said, "I kin do it. But I haven't got the heart to strain the gun."

Lafe Allen and Heber Schuster were among the early-comers to the new colony. In the vicinity of Deep Springs they had been cabin-mates so long that they had had a row, and had set up independent housekeeping, each in his own end of the shack. Thus they had gone on dwelling, refusing to speak. When word of the new prospect percolated to Deep Springs, both Lafe and Heber wanted to set out but each was afraid the other would be pleased if he packed up. It would look like surrender of a point of honor.

Finally some telepathy between the two hostile cronies

brought decision. No words were spoken but the burros were rounded up and packsacks were slung. Arrived at Panamint, the pair pretended to look about for separate quarters, found none, and still on wordless terms began to roll up stones for walls. When the dwelling was completed it had two fireplaces, Lafe doing his cooking at one end, Heber at the other—just as if they hadn't moved a hundred miles.

A rider comes into San Bernardino after an upland trip and brings word that the new district is wild with enthusiasm. At Los Angeles the *Star* is noting, "The excitement in San Francisco about the Panamint mines is almost equal to that about White Pine in the Fall of '68 and Spring of '69."

Up the canyon toils Jacob Cohn, sturdy member of an ancient mercantile race. He looks about with squirrel-like eyes, goes back to Fort Independence and closes up his little post store, continues on to Los Angeles for a stock of hats, blankets, overalls, shooting irons and cutlery, and is back at Panamint as fast as a man can cover a five-hundred-mile circuit and accelerate a string of burros.

Up the canyon come others of his stripe—Isaac Harris of Harris & Rhine, over from Independence to lay the beginnings of what will be the town's chief emporium; little Mrs. Zobelein, to be on hand in case the camp acquires ladies and the ladies require hats, ribbons, gingham; and presently a character who hasn't missed a new mining ruckus since the intermountain country opened. Uncle "Billy Bedamned" Wolsesberger, aged peddler, limps the entire 417 miles from Eureka driving his stock of goods before him on a little donkey.

There are others who arrive after pushing wheelbarrows across the deserts, that midsummer of '74; and John Schober, who hears that the camp has plenty of knotty trees but no lumber, comes on foot from Bakersfield carrying his whole mill—an immense hand whipsaw—on his back.

Up the canyon comes Captain I. G. Messec, a long-bearded Georgian who crossed the deserts in '49, and became a freight packer to the Trinity country in the '50's; who served as sheriff of that rugged country and as captain of its volunteer forces against the Klamath Indians. Once while taking a mule train through, Captain Messec encountered a stranger in the snow. The stranger was just about done for, but he gasped out a story which the captain took to Scott's Bar along with the injured man. Gold had been discovered on the Klamath! The result was the rush to Yreka and the Oregon border. Ever since, Captain Messec has been looking for another Yreka; and if rumors are to be credited, Panamint will do . . .

Up rides R. E. Arick, who claims to be onetime mayor of Virginia City, and who concludes that here is a second Comstock fully worthy of his political talents. Up rides primly respectable Miss Delia Donoghue in a light wagon that is piled with stove, pots, pans and comestibles. Miss Donoghue, who knows the age-old way to men's hearts, is determined that the citizens of Panamint for all their wild ways shall have home cooking. Up the canyon comes Charles King, once sheriff of Placer, to set up a meat market. He brings his equipment in a two-wheeled cart that shall see service of a sort unguessed by its proprietor. Up presses Charles G. Meyer, one of the veteran assayers of the Coast, who came to Eldorado on the first overland stage and has been through all the principal mining frenzies; he bears now the commission of Senator Jones to appraise the ore of Surprise Valley.

The zest for discovery is not confined to Surprise Valley's azurite and malachite cliffs. One locator halts on the way and stakes out Coso's craters for his private sulphur mine.

Another enthusiast goes down Surprise Canyon to its outlet, turns north, and stakes out for himself a square mile and a quarter of Panamint Valley—of which there is plenty— and announces himself the proprietor of salt for the reduction works and enough left over to supply all the tables in the world. When it is pointed out that he is competing with the Pacific Ocean, a notable salt-producer, he looks over his mile and a quarter, guesses again about its depth, and announces that big as the competition is he can still lick it.

From neighboring canyons come reports of discoveries of argentiferous galena, silver in rich black chlorides, silver in roasting ores and ores free-milling; until proud Virginia City, with the enormous jewel of the Big Bonanza secreted in her breast, is constrained to take notice of this distant show. "The ores found at Panamint," growls the *Territorial Enterprise*, "are nearly all more or less base . . . Wherever the vein penetrates a mass of slate, as it frequently does, the quartz is invariably barren . . . The new mine is merely another opening in the Base-metal Range."

Many have set out from Pioche along the path taken by Dave Neagle, though the season is malignant along the 280-mile way. Others are thrusting southward from central Nevada. Such a party out of Eureka, composed of Bark Ashim and five others, start for Panamint in August. Matters go well until, a hundred miles from their journey's end, the party runs out of water. They are in a region of saline valleys and barren mountain mazes, and in searching the gullies for springs become hopelessly separated.

Five of the party find the return track to their halting-place at Lida, where they arrive more dead than alive. Ashim has been left behind. He is too far gone to run when a small band of Panamint Indians discover him. They take him to their camp and revive him with cautious sips of water and a meal of gourds. Not until he is set once more on the trail for Lida does Ashim know whether he is being treated hospitably or being prepared for some ceremonial.

More than one lost traveler pays the greater price. William Wilson, an old Indian fighter and prospector who founded the Pinegrove district in '66, is found dead beside some springs in Panamint Valley. He was but a few miles from his goal when he came upon the pool, but his ebbing strength was not able to survive the disappointment of finding it hot and sulphurous.

Fifty miles away the stage tenders at Indian Wells come upon the body of a man who has crawled to within two miles of their station. The body is lying in the sagebrush, a hat down over its eyes as if to shield them from the sun; over one arm is looped a bloodstained bridle, and nearby is an empty canteen and a revolver with one chamber discharged. The story is plain: the traveler grew desert-crazed and lost or shot his horse, continued on afoot and gave out when his canteen emptied.

Farther north three men set out on horseback from Independence to hunt up an asserted lead mine in the Panamints near the head of Death Valley. The mine was remarked three years previously by members of Lieutenant Wheeler's government exploring column. After tarrying a few days in the northern Panamints, unsuccessfully looking for this deposit, one of the present trio, V. A. Gregg, leaves his comrades and strikes out alone for Panamint.

As the month is July, and his course lies through the furnace-like length of Panamint Valley, this is dangerous procedure. Panamint is seventy miles onward and there is not a drop of water to be had before reaching the springs just short of Surprise Canyon. The occasional wind-puffs that come up Panamint Valley's salty floor, Gregg says later, were more blistering than still air. His canteen "seemed to fill up with steam" when he removed the stopper.

Gregg pushed on down the long shadeless plain hour after hour, mounting and descending the slopes which fanned out from the succession of alpine gorges on his left. One barranca proved of unusual depth. Man and horse proceeded cautiously over rolling rocks.

As they crossed the bottom and set to the opposite ascent, Gregg's eye fell on a disconcerting object. It was the bony framework of a human leg, held together by toughly dried ligaments. The relic was probably Egan's, the second of two guides lost while the Wheeler party was looking for springs and passes in Death Valley's mountains. Apparently the guide had died in the Panamint heights and had been carried out into the desert by subsequent rushes of storm waters. Gregg hurried from the gruesome relic, making for the warm springs and there setting up camp. He had had enough for one day.

Nor was blistering heat the only weapon which the desert turned against its traversers. James Clancy was two or three months behind his fellows who had made the crossing from Pioche. He neglected to bring sufficient blankets, which seems like an excusable error in those precincts. The sands turned cold and Clancy reached Panamint town just in time to be laid away, first permanent resident of the side ravine thereafter known as Sour Dough Canyon.

As these characters ascended Surprise Canyon, they encountered another coming down. Bob Stewart's shovel and bucket rode atop his burro, and he strode beside that faithful animal with the unhurried gait of one who was off once more on the journey of decades. "If a man looks for a thing for twenty years, turning over every rock, and finally finds what he's looking for, folks call him a 'lucky prospector.' Well, I've been lucky. So I'll just be pushin' on, boys— got another twenty years' searchin' to do."

Out there on the deserts you'll encounter him still, or his undying type, though by now his burro is gone to more cheerful pastures and six cylinders with high clearance have given the gray old pilgrim new command of Distance. You'll encounter him by the mesquite spring and you'll en- counter him in his cabin, and he'll receive you courteously and fill you up hospitably with coffee that will turn your fillings black and yarns that will do things to your hair. Hail to him, sagebrusher and old-timer, stout-hearted out- o'-luck and transient millionaire! May he dwell on till the knoll behind his shack reveals the Gunsight and the greasewood at his doorstep puts forth roses.

Two days after Raines had arrived in Los Angeles, that previous December, a terrific duststorm had struck the up- lands. It came as a thick yellow mist from the southeast. Stages trying to ascend Tehachapi Canyon above Bakersfield found it utterly impossible to face the wind. Lighter vehicles were overturned like baskets. During this wild simoom a number of herders and several thousand sheep had been caught on winter pastures in the Coso Mountains.

Now, in sudden unseasonable chill of ensuing autumn, a wayfarer for Panamint was caught in the same bare hills. After being lost for some days, and saving his feet from freezing at night by burying them in the sand, John Carlin pulled himself to a deserted cabin near Black Springs. He kept himself alive by devouring some pieces of tallow candle. Other passers on the way for Panamint heard speech issuing from the cabin, and entering in curiosity found Carlin holding maniacal conversation with the bones of the sheepherders who had perished in the '73 dust-blizzard.

Among those who started out from Pioche for Panamint were three young pedestrians—William Honan, Peter Dawson and Charles Olsen. They reached a point fifteen miles beyond Hiko in the Silver Mountains. In their minds there had begun to grow some doubt as to whether a fortune was really waiting at the rainbow's end. Their provisions and shoes were used up and they concluded to retrace to Hiko Valley and look for work.

On their way they fell in with an Indian, Tempiute Bill, and eight of his braves. Though the youths considered their equipment sorry, the Tempiutes adjudged it elegant and desirable. They dropped in behind the trio and walked in their tracks.

Honan was the first to grow suspicious. He lengthened his stride, signing to his companions to keep the new pace. But they had been picking 'em up and putting 'em down for too many miles to feel like hurrying now.

The gap between Honan and his friends increased. It widened still more when Honan began sprinting. A rifle ball spat after him, catching him in the shoulder. Thereupon he set to throwing up gravel in earnest. A flurry of arrows fell short, but out of the corner of his eye he perceived that they had not fallen short of his companions. With that he put all he had into his pumping legs and kept at it. When he reached Hiko and a party of whites went out to verify his tale, they found the body of Dawson pounded and half buried by rocks, and after a day's search came similarly upon Olsen. The aftermath was a general mêlée between ranchmen and Indians, a massacre of a native village, and the hanging of Tempiute Bill and one of his followers from a window of Hiko's only two-story building.

The same week that saw the slaughter of the Pioche youths found Patrick Lyons of Benton heeding the call to Panamint and packing up his burro. Three miles down the road he stopped at the cabin of two acquaintances and urged them to come along. Messrs. Forey and Lavell, who owned the cabin, declined. They had no great opinion of the future of Panamint. At this aspersion of his newly adopted town, which he hadn't seen yet, Lyons started to do some fancy scroll work on Forey with a knife. Forey's retort was a pistol ball which ripped Lyons' lung.

Deciding that a man who thought so much of Panamint must after all be right, Forey and Lavell disposed Lyons comfortably in the cabin and hit the trail in his stead. It was quite a few weeks before Pat Lyons reached Panamint and rejoined the pair. Though loser in the physical argument, it no doubt pleased him to find that he had won the verbal.

Those arriving at the top of the canyon continued to tell of desert roads filled with teams and people on horseback. Independence, eighty miles away in a crow's line, was virtually emptied of its slender population, its weekly *Independent* styling it "the deserted village, as everybody is off for Panamint."

The rising camp, possessed of several saloons, a store or two, and a cemetery, now went in for another municipal adornment—a justice of the peace. William C. Smith, an understanding young man, was elevated to that office in

July. His services were to be in continual demand but his duties simple, consisting mainly in finding "the homicide justified on the grounds of self-defense." In the subsequent career of Panamint, despite numerous affrays, any other judgment was rarely voiced by the accommodating justice.

By the middle of September there were three or four hundred citizens in the hundred-yard-wide swale where, a few months before, nothing had been audible but the wind in the pines and the babble of a nameless brook.

Off in Pioche, Bob Archer of the Sazerac Saloon was showing his friends a letter from Dave Neagle:

"I seldom advise parties to come to new mining camps, but in this instance I have no hesitancy in giving you all the information you desire, and advice also. I consider this the richest mining camp on the Coast . . . There is not a man in camp idle that wants work, and more men are wanted. I will, in a few days, have the finest saloon, out of San Francisco, on the Coast, 20 X 50 feet, and the finest bar and stock that money can purchase. Times are lively, money is plenty, and everyone is in high hopes and full of confidence. The population is increasing every day. We will have 2,000 or 3,000 men here this winter; so you can judge for yourself what our camp will be. The above is my honest impression of the camp—a second Washoe, a place where work is plenty for all who come."

The Austin *Reveille* watched the developing restlessness among its subscribers and cautioned them that it was "better to bear the ills they have" than to "undertake a weary journey over mountain and desert at this season of the year, which cannot be done without great privation and hardship." But it was bound to admit that "this warning will have no effect, however, on those restless spirits who can subsist for nine days on the smell of a dish-rag, when on the trail leading to a new camp. They would start at any season, even if they had no better outfit than three soda crackers, a bunch of matches, a jackknife and a half a horse blanket, and they knew there wasn't a drop of water for a hundred miles."

"Anything but flattering reports are received here from Panamint," railed the Pioche *Record,* inconsistently adding: "The fever is still raging. A great many Piochers have already gone and more are preparing to go. William Culverwell, of the Bullionville stage line, starts with two stageloads on the 6th for Panamint, fare $40. He also starts a freight train on the eighth."

Though there were rough characters galore streaming for the latest boom camp, and Panamint's name for years after was to be synonymous with deviltry, its early reputation for good behavior seems to have been maintained at the moment. "An Orderly Town.—No new camp has had less of rowing, cutting and shooting than has Panamint," marveled the county weekly at Independence. "There is but little drunkenness, and the inhabitants seem to be of an intelligent, thrifty class."

# 11

# ONE GOOD SENATOR DESERVES ANOTHER

I⊤ was at July's end that four assorted travelers took day
coach at Oakland, sleeper at Lathrop, stagecoach the next
morning at Delano, and after a substantial breakfast in Monsieur Escalet's hotel in Bakersfield boarded a leather-swung
"mud wagon" behind six horses for the crossing of the Sierra
Nevada. Three days' bouncing, in cushions over the mountains and saddles over the desert, landed them at the top
of Surprise Canyon.

With two of them, "Colonel" Raines and Harry A.
Jones, brother of the well-endowed Senator, the camp was
acquainted. A third member of the party, Trenor W. Park,
wealthy, aristocratic and fifty-one, had been attorney for San
Francisco's Vigilantes in the gold-rush Fifties. Since his return to the Atlantic he had become a banker, director and
counsel of the Pacific Mail Steamship Company, and as president of the Panama Railroad one of the richest men in the
country. Prospect of large ore freights from Panamint to
the British Isles, with boundless revenues for his isthmian
rails, had brought the Vermonter into the Surprise Valley
adventure.

The fourth of the arrivals was an Ajax of a man who
seemed tremendously at home as he swung from stirrup and
planted foot on the rough stones of a mining camp.

At Washington, William Morris Stewart dwelt in an

imposing three-story pile complete with conical tower, porte-cochère, bays, walnut furniture, speaking tubes, three bath-rooms, and the hugest pier-glass mirror in America—a man-sion whose bric-à-brac gewgaws and startling dormers must have represented the ultimate paroxysm in the world's most painful period of inside and outside decoration.

But Stewart Castle on Dupont Circle, with its stable that overshadowed the British embassy on the rear street; with its metal lightning-rod and shiny weather vane at the top of its cupola; with its grand staircase, its tapestried ballroom, and its chandeliers of carved cupids holding gas jets in their feet, now was all very far away. Here—cliffs, sagebrush, pines, the scars of recent storm, the hastily propped-up Hotel de Bum, the grinning miners, the rough resorts—was reality.

The Bill Stewart who tossed the lines over his horse's head, loosened its cinch and turned to invade the canvas inn was six feet two, looked taller, and was at this time forty-seven years of age. His frame was solid. His bearing was commanding. The native self-confidence of the man who, before reaching the national forum, had grappled and tossed the most aggressive intellects of his day in the rowdiest arena on earth—the legal circus of the Comstock—was as obvious as the whiskers on his face. And those whiskers, whether in their present leonine buff or their later states-manlike white, were the noblest that flew from any chin between Potomac and Chesapeake or between the Sierra Nevada and the Great Salt Lake. John P. Jones, his col-league, streamed a manly beard and it was rated by some the second-finest in or out of the intermountain country, but compared with Stewart's oriflamme it was a modest banner.

Everything that Bill Stewart undertook went with that kind of vigor. There was a quality about him that made things crackle and pop and happen. The New York farm boy and wood-chopper who had elbowed his way into Yale College, the Yale sophomore who flung books aside to join the '49 gold rush, the youth who crammed aboard a shaky steamship with fourteen hundred other human mackerel, to arrive at Panama with the vessel's engine broken and all masts overboard; the boy who shouldered his way across the Isthmus and aboard the little propeller "Carolina" to arrive at the Mother Lode racked with fever and to lie for days under a tree, mending himself by will-power and starvation; the tenderfoot who learned to whack bull teams up the long grade between Marysville and Grass Valley; the tall, light-haired miner in overalls, flannel shirt, white felt hat and coarse, sockless shoes who sank pick into Buckeye Hill and cracked open a treasury—these had been the successive Bill Stewarts of youth's rushing cyclorama. When he went into a law office in Nevada City, California, he left behind him as practical accomplishments the camp's first sawmill, a ditch and flume winding for miles along the mountainside, and the West's first quartz miners' code.

Lawyer McConnell, reigning Blackstone of the region, provided the likely young fellow with a corner to study in. Stewart was district attorney soon, in another two years state's attorney-general. Then he was back in noisy private practice that oscillated between the mountain towns of the Yuba.

To this rustic setting the glamor of the times had attracted an exceptionally brilliant bar. Its ornaments had come from all parts of the Union, with men from the South slightly more numerous and belligerent. Arguments that began in court were continued in barroom and back room, and appeals were frequently taken to the field of honor. From these tutors Stewart took repeated legal trouncings, but on the grindstone of their wisdom he grew in sharpness and it was presently discoverable that while his adversaries often had the law, Bill Stewart had the jurors.

As he rose in prestige and estate Stewart built a house, the most pretentious in Nevada City, and informed his fellow-townsmen that he was about to go down to San Francisco and bring back the finest girl in the Golden West. But he had too soon disclosed his brief. A rival attorney slipped down to the Bay ahead of him, and when Stewart presented his case he found that his long, hard study of men had taught him exactly nothing about women. The defeat was complete and there was no appeal. Stewart philosophically got himself to a dramshop.

In its friendly atmosphere, testimony goes, he found an acquaintance. It was ex-Governor Henry S. Foote of Mississippi, later his law partner, who now invited the non-suited barrister to have a drink. Stewart accepted and proposed another, and so passed a day, a night and part of another day.

It was while they were lying side by side on the floor of Foote's hotel room that the ex-Governor said:

"Stewart, you are a damned Northerner and your political principles are a disgrace, but personally you have qualities and it would be a pleasure to me to befriend you, suh!"

"You can do me a service right now," accepted Stewart, who had been thinking about that house in Nevada City. "You can let me court your daughter Annie for my wife."

"Well," considered the ex-Governor, "as I remarked before, I can't understand your vicious Northern doctrines, but if Annie doesn't mind I'll not go back on my word. She might do worse."

It took a week for Stewart to make himself a bridegroom, but he was back in Nevada City and his mansion had a chatelaine on schedule. They were a devoted couple, and after nearly half a century of married life he would speak of her passing as "one of the tragedies of my life."

The unfolding glories of the Comstock just across the mountains offered the perfect field for an opportunist of Stewart's caliber. The tremendous riches there compressed within a few overlapping claims, the debatable ownership, and the general pandemonium and anarchy called him as a trumpet.

He arrived in Virginia City in March, 1860, and found men by scores who had lived through bitter winter in tents, behind rocks, in holes in the ground or wherever they could find shelter from the smiting "Washoe zephyrs."

High against Mount Davidson, sharply etched in the transparent air, the roofs and tents of the throbbing camp could be seen from Carson Desert for sixty miles. Virginia City lay halfway up the granite mountain, pitched at a steep angle, the shacks about the Ophir discovery mine being connected by trail over a shoulder of the peak with Gold Hill at the top of a canyon two or three miles distant and Silver City down-canyon beyond. Stewart was soon to see that Indian trail a solid street of buildings completely connected, and situated over a deep, complex chain of underground works almost equally connected; and in neither the galleries beneath the earth nor in the houses above would men know night from day.

Nevada was then "Western Utah." Its seat of government, if any, was at Salt Lake City, which was as distant in time as Asia or Panama today. But far as the miners were from the Latter Day Saints in geography, they were farther yet in spirit. They knew nothing of Utah laws, distrusted Mormon authority and organized their corporations under California statutes with their headquarters and stock-selling activities centered at San Francisco. The confusion caused by miners applying to deep workings the California principle of pursuing quartz ore through all its "dips, spurs and angles" was already resulting in some bloody collisions.

Highly informal arbitration was often invoked. Stewart was an arbiter on one typical occasion. The setting was a small back room. Stewart faced the door. It opened to admit Sam Brown, a self-cocking bad man who boasted sixteen notches to his gun and a hatred for tranquillity.

Sam shouted his intention to take charge over the proceedings. Stewart met him with two snub-nosed derringers that had little range, but the power within that range to launch solid one-ounce balls.

Red, liquor-inflamed eyes looked into blue and it was the red that fell.

"I like your kind," mumbled Sam. "Come 'n' have a drink."

Over the bar they cemented friendship and the curly wolf retained Bill Stewart to represent him in a lawsuit that was shaping in Aurora. Sam Brown was Stewart's client, however, for but a few hours. He was slain that night by an offended citizen with a shotgun.

Two days before Stewart reached the Comstock, David S. Terry, an old adversary and a grim one, had arrived from the other side of the mountains. Terry was a man of intensity. He had nearly been hanged in the Vigilante rising of '56 though chief justice of the state supreme court at the time, and he had recently slain Senator Broderick in a duel.

Terry conceived it his duty to seize and fortify a portion of the southern or Silver City end of the Comstock for his clients. It was good, sound, nine-point law and he acted on it. Bill Stewart was his adversary in the case that promptly shaped itself. While quite prepared to lead an armed charge of his own on this treasure-laden Bunker Hill, Stewart finally talked his fiery opponent into adjourning matters to Carson City, the territorial capital, and settling them by the warfare of writ and injunction.

The situation was complicated by the presence of two judges, each claiming federal appointment. Stewart and Terry agreed to accept one jurist. Before the sun was up next morning, the other judge, suspected of belonging to the Terry camp, had repudiated the agreement and was on the street proclaiming jurisdiction.

"I belted on my pistols," recalled Stewart later, "and started down town, seeking Judge Flennicken. I met him on the square, now occupied by the State House, in front of Pete Hopkins' saloon. 'Good morning,' he said. 'Good morning.' 'What's the news?' 'Bad news, indeed,' I said. 'They are slandering you. They say you are claiming to be judge and defying the authority of Judge Cradlebaugh.'

"I told him I had anticipated that something might go wrong, and had taken the precaution to be deputized by Marshal Blackburn to summon a posse to assist in executing the orders of Judge Cradlebaugh, and that I summoned him to carry a musket in front of me . . . He raised his hands imploringly, saying:

" 'Is there no way to avert it?' 'Yes, if you will do as I say,' I replied. He consented by not resisting, and I took him by the coat collar . . ." Whereupon followed a curious scene, the judge being hustled into the telegraph office and required to send messages to all parties concerned, announcing his entire disinterest and disqualification.

Other cases brought other judges, but not always peace to the Comstock and not always righteousness. It is related that one of these Solomons, a notorious chief justice, was so entirely businesslike with litigants that cases simply could not be tried until justice had been oiled. Price for a favorable decision was set at $10,000 in one instance, the decision to be returned when court convened next morning. There was some difficulty getting the coin together but at length it was ready in a canvas bag. The hour being late and the jurist deep in dreams, the knock at his door was answered by the mistress of the house in her nightgown.

"Is the chief justice in?" asked a whisper out of the night.

"Yes, but he's asleep."

"I've brought the money."

"I'll take it."

The courier handed over his bulky bag and the judge's lady made an apron of her nightie. There were fifty pounds in eagles and double eagles and they tore the garment completely off. The courier drew shut the door.

As the victorious attorney in that era of terrific legal give-and-take, Stewart received fees running up to $200,000 a year, and he doubtless earned them. Tossed into enterprises that proved successful, they pyramided.

Nature's forces were less amenable to his burly will. When he had half a million laid out in mills and mines, winter charged down the slopes of Mount Davidson in the shape of a deluge, destroyed the result of his labors lock, stock and barrel, and went roaring on down Carson Valley and off into Carson Sink.

Whereupon the lawyer and mill-owner became horseback rider and pedestrian, scaled the Sierras through blizzard by sighting from tree to tree, dodged avalanching snows, found inland California a lake, swam, rowed and sailed until he reached San Francisco, borrowed from a friend to hide his bankruptcy and returned by heroic journeying to the scene of his overthrow before anybody knew he was broke. That was a journey which few other men dared to undertake. On his way back, with the credit for $32,000 in his pocket, he found the same travelers marooned at Yank's Station on the top of the Sierras that he had bunked with for a night the week before. Indomitably Stewart built his fortune up again.

When Lincoln was running for his second term and the new battle-born State had also to select a governor, a congressman, all county officers, a legislature and two senators, Bill Stewart decided to be one of the latter. The campaign was a rousing one. The Lincoln men equipped great freight-wagons with platforms, beds and seats, and with a rostrum in the center on which stood a barrel of Kentucky rye whisky with a cup chained to it. Each wagon was drawn by eight to a dozen mules. Accompanied by a cannon on wheels, plenty of powder, some Indians, a quartet of singers and a rawhide band, this caravan set out.

Through dust and uproar one sees upon those rocking freight-wagons Professor Silliman, the venerated Yale geologist; Governor Nye, the other Lincolnian candidate for Senator; Frank Ganahl, Bill Claggett and other orators of the place and day; and Bill Stewart the Samsonian figure of them all. By means of much tumult, much speaking and much black-barrel firewater, the treasure-casket of the Union was snatched to the Republican bosom.

At Washington, Bill Stewart proved as usual to be a

tremendous worker. When Nevada took its position as the thirty-ninth state, its silver lodes spilling richness like a broken wine-keg, the elder states were hopeful that taxation of the claim-owners might pay all the expenses of federal government. Stewart waged a battle for the possessory rights of the claim-holders; his hand shaped what became the mining law of the land. He also drew up the Fifteenth Amendment, guiding it through parliamentary complexity though the job kept him for forty consecutive hours in his seat, and the words in which he framed it—"The right of citizens of the United States to vote and hold office shall not be denied or abridged by the United States or any State on account of race, color or previous condition of servitude"—are close to the form which it holds in the Constitution today. Lover of action, he declined Grant's offer of an appointment to the Supreme Court.

It was to Stewart in Washington that a slouching individual once came and, on the strength of old Washoe acquaintanceship, asked for a grubstake while he wrote a book. Stewart gave the shabby author a cubby-hole, waved him to the cigars, and there Mark Twain wrote "Innocents Abroad."

The worst rout of Stewart's political life was very fresh in his memory as he stood in Panamint's rough street this July day of '74.

It had recently been engineered by John Sherman, developer and introducer of the Mint Bill of the previous session. Throughout his drafting and presentation of his pet measure, Sherman had dwelt on the glories and beauties of the silver trade dollar as used in dealings with the Orient and Stewart of Nevada had been led to be one of his supporters, even making a speech in favor of the bill. But when it became law and Stewart found time to read it, he discovered that the standard silver dollar as used in the United States had been dropped from the list of coins.

Demonetization, a European invention raised like a dike against the menacing tide of silver, had crossed the Atlantic and stood installed—with Bill Stewart's help—as the policy of America.

Sherman's management of this trick was quite of a piece with the school of politics that had graduated Stewart himself, but in this instance it spelled debacle for the Nevadan. Silver made Nevada, and when Stewart's constituents found that he had voted for the "Crime of '73," all his explanations were received with hoots.

The rout had come at a particularly unwelcome time. The end of his second term was approaching. Stewart liked Washington and doted on that big wigwam of bays, towers and dormers on Dupont Circle to which he was just putting finishing touches. Visitors to the national capital hurried to view it as they would hurry in later years to view the Washington Monument, neglecting the White House itself in their desire to see and exclaim over the sharp windows, steeple and columned porticos of Stewart Castle; to admire its stained glass windows and stand awestruck before its circular vestibule whose double row of gas jets, on winter evenings, cast a corona that rivaled that of the Senate's own galleries.

"But the wonder of it all," breathed a contemporary rhapsodist in one of the eastern journals, "is the parlor. To get its full effect we should step into the hall, where all is massive and heavy and dark, and drawing back the great doors, that slide so easily a child could push them, cross the threshold from the subdued reality of walnut and morocco to the glitter and glory of fairy land.

Courtesy of Roy Jones.

To Mrs Finly
Wm Stewart
Nevada

Senator Bill Stewart's last home in the desert. Bullfrog, Nevada, 1935.

Through these windows Bill Stewart could see the
purple Panamints. Ruins of his house at Bullfrog,
Nevada, 1935.

"There has never been anything like it seen here, and seldom anywhere else. Up to the very edge of the ceiling gilt cornices rise, carved into most exquisite devices, from which superb curtains, woven by the historic Gobelin looms, fall in luxurious folds to the carpeted floor; chairs magnificent enough for thrones; graceful tête-à-têtes, wonderful sofas and divans, of gold and Aubusson tapestry . . . ivory, Parian marble, alabaster, bronze, gilt—everything that art can conceive and skill fashion."

The pile could not have been more terrifically the capstone of its period, one gathers at a historical distance, if General Grant and Queen Victoria themselves had collaborated to conceive it.

Stewart Castle was already the scene of gay charades, tableaux and balls; and Bill Stewart and particularly his family would have liked to live in it forever. But his elaborate scale of living and some unlucky speculations had the Nevadan badly bent. An unfortunate lawsuit was also shaping against him over a mining incident in Utah. And now, this political blunder. Moreover, up and down the Washoe hills William Sharon, beaten by Jones for the other toga at the last election but not subdued, was now demanding the classic garment from Stewart, and Sharon had millions to demand it with. Stewart realized with regret that Sharon was going to get it.

Thus it befell, at an acute juncture of his affairs, that J. P. Jones had suggested the Panamint partnership.

One senses the striking similarities and differences of the two colleagues as W. M. Stewart and J. P. Jones settled into Stewart Castle's walnut-and-morocco to talk that proposition over.

Jones was a thinker and logician without a peer; a man of

little schooling, but of wide reading and of a genuine capacity to broaden and deepen. Stewart as a youth had reached the middle of his second year under New Haven's elms, and could claim a shade the better formal education. In spite of his current downfall, he was also reckoned the better politician, but this is to be doubted; both men definitely had the common touch. Jones was the bitter foeman of all railroad monopolists, though equally the foeman of audacious Sutro who by his Comstock tunnel was out to break the grip of cliques on the Washoe lode. Stewart on the other hand was a pronounced supporter of the Central and Southern Pacific railroad barons and was one of the directors of Sutro's tunnel enterprise.

But there their differences ended. The pair were of an age and both had the log-cabin background. Both were Forty-niners. Both had been sturdy with the pick. Both had moved on from the Mother Lode to the Washoes. If Jones had become a successful Comstock superintendent, tracing and extracting millions, Stewart had made himself the highest paid lawyer on the lode and at times its dictator.

Both were indefatigable. Both had acquired a wisdom not contained in libraries. Both venerated silver. Both would serve, in the ultimate, three decades in the Senate. Both loved a gamble, a joke, or an adventure. Both had won and lost fortunes enough to suit a dozen men and both would lose and win more.

William M. Stewart, with Engineer Stetefeldt's report in his hand, had looked afar from the deep leather comfort of Stewart Castle to the dim crags of the Panamint Mountains on the other side of the continent. Crags that lifted up from relentless sands to implacable skies, holding that beacon to men who saw in its long rays the silver of the Gunsight, the yellow of the Breyfogle, the insistency and mockery of John Goller's dream, and all those figments and illusions of twenty-five years' searching. But this time the illusions seemed resolving to something solid. John P. Jones had his brother Harry Jones' report on his knee. For Jones, the checks were down.

Bill Stewart's light blue eyes, before they came down from the crags, may have seen more than met John P. Jones' dark brown.

At any rate, Stewart nodded.

One more lame-duck session at Washington and he would be ready to give the enterprise his heartiest personal attention. For further capital they would bring in Trenor Park. Stewart admittedly could do with another rendezvous with fortune. Out there where the sagebrush met the porphyry mountains he had never failed to find her, and he would find her anew.

The master of Stewart Castle, having swung now from his horse, strode over stones that had lately been the channel for a gully-washer and turned in at the doorway of Panamint's canvas inn. At his open-handed signal the inhabitants of the camp trooped forward.

For a moment he stood in that doorway, his glance reviewing the buffeted townsite and its expectant denizens. Virginia City in its first sorry, gorgeous days had looked like this.

From the foreground his eye swept on, ranging up and up those slopes to heights that danced in the sun.

Here was air a man could breathe.

## 12

## THE COMING OF THE FREIGHT TEAMS

STOREKEEPER MEYERSTEIN was checking over his stock and making up his fall order for boots, shirts, saddles, canned goods and sombreros.

It was September, and San Bernardino can be sweltery in September. Down the road Blacksmith McCall could be heard tap-tapping on his anvil. The desultory clingclang sang a story of cherry-red iron and smoking hair; it gave voice and odor to the summer heat. But Meyerstein, pencil in one hand, wet bandana in the other and glasses on his forehead, bent to his task. Showers would be cooling the overhead desert in a few weeks and prospectors would be outfitting. They would be setting out for the Hassayampa country in Arizona and for the Chuckawallas and the Bullions this side of the Colorado River; they would be climbing Cajón Pass and picking up the endless search for lost leads out on the Mohave. But above all, what with this new to-do, they would be outfitting for Panamint.

There were no railroads, no telegraphs serving San Bernardino. A storekeeper must get his orders into San Francisco early if he wanted to be ready for seasonal business. So many pairs of cotton pants, so many pans and flasks, so much ammunition—

Meyerstein's doorway, a glaring rectangle of white roadway and sky, was just then filled up by an entering figure. Facing the dark interior, with all that light at his back, the visitor was pretty much an opaque silhouette. Nevertheless Meyerstein, as he brought the spectacles down to his nose, sensed that this was a beard and frame at the moment unfamiliar.

The stranger was forthright. Would Meyerstein care to figure on five hundred tons of assorted merchandise, as per list submitted?

The horse and saddle visible at the hitchrack were of a quality that tended to establish credit. The stranger also had an air about him which furnished its own credit rating. Yes, Meyerstein & Company would be delighted to supply, given reasonable time, five hundred pounds of anything. Wagon or packmule delivery could be managed. Calicos, nails, shovels, whisky—what did the customer require?

Five hundred tons, not pounds, was the item under consideration. The merchant would kindly look over the schedule. There would also be 163,000 feet of lumber, transportation and found for two hundred Chinamen, and a thrice-a-week pony express.

When Meyerstein finished studying that list and perceived that he was really hearing about tons, not pounds, and Chinamen by the hundreds and skinners by the shoal, he went down and saw Blacksmith McCall. He left McCall leaning dizzily against his forge, wondering if the sun had got him at last. Meyerstein, who never joked, had asked him how quickly he could bend four shoes apiece to the hoofs of a hundred mules.

Bill Stewart was moving.

Up through Cajón Pass and northward sprang the pony of the new lightning service on a morning of mid-September.

With a relief waiting at Black's Ranch ninety miles onward and relays of fresh mounts at scattered intervals, the new service was expected to make the flight to Panamint between sunrise and sunrise.

A few weeks later the first batch of forty-five Chinese laborers arrived by water from San Francisco. They were put in charge of Charley Craw, one of Meyerstein's hastily drummed-up lieutenants, and told to keep going with shovels and wheelbarrows until they had filled every wash between the Cajón Pass and Panamint. Charley Craw went to work with a will, being especially mindful to keep his eye open for waterholes, and

"Up through Cajón Pass sprang the pony."
Advertisement in Los Angeles *Express*, after the initial rate had been cut.

seventy-five miles onward he was rewarded by a fine spring, in a ravine under Pilot Knob, interrupting the longest sterile stretch on the pathway.

In the same month the steamer "Orizaba" put in at Wilmington with a towering engine and boiler for Panamint and bearing also the second installment of Chinese, which Stewart had found time to recruit beside the Golden Gate. Before October was out, Meyerstein's Panamint Freighting Company had two hundred tons of goods moving over the deserts and Newt Noble had resigned his job as sheriff of the biggest county in the U.S.A. in order to command the teaming enterprise. Just now he was scouring the county for every wagon, loose mule and sober skinner in its length and breadth.

By the end of November merchant-contractor Meyerstein was handling three hundred tons of mixed freight a month, 1500 tons more were en route by sea or piling up near Los Angeles, and Noble was hunting for additional rolling equipment at points as far away as Yuma, Prescott and San Diego. A visitor on his way to have a look at Panamint a little while before had passed through a San Bernardino that was a village sleeping in the sun. Returning a few weeks later, he found the gateway hamlet completely transformed. Every anvil was clanking. A hundred buildings were going up. Stations were being erected beyond Cajón Pass at eleven water-holes.

Other routes were likewise humming. In August, Stewart, Park and Harry Jones had passed through Bakersfield. Their advent at the moment had meant so little to the village that editor-printer Joe Acklin had remarked in his *Southern Californian:* "We were too busy in our office to interview the gentlemen." But now Bakersfield had waked to its strategic situation. The road beyond Indian Wells, whence freights from Bakersfield moved out on their way to Panamint, was reported to be "a continuous line of teams all the way along."

It was heavy going out there, the sands and grades having led Stewart to turn to the San Bernardino route as an alternative. An army of draft animals, hauling a five-stamp mill for R. L. Jacobs, came to grief east of Indian Wells and the load had to be left for days while drifts of sand closed over and men dug and swore. When the wagons were finally got out, it was only to sink to the axles again in Salt Wells Valley, a region of squidgy, soapy marsh.

"After this is passed," said an eye witness, "comes a mountain, the descent of which is at an angle of 45°; at all other times is heavy sand, so we infer that the teamsters don't whistle and sing all day from sheer happiness."

Reports that Meyerstein's teams out of San Bernardino were also finding desperate going, with "water in but one or two places in a stretch of ninety miles, and sand so heavy that the brake-blocks drag the ground long ways at a time" encouraged Remi Nadeau to grab for the trade. His Cerro Gordo route lay between the two. Nadeau was operating fifty-six fast-freight teams of twelve and fourteen mules each, continually on the go; they were carrying down four hundred bars of Cerro Gordo lead-silver bullion daily. If a wagon broke down, the eighty and hundred-pound "pigs" were casually tossed beside the road to wait for another trip. Nadeau's roundtrip journeys from Owens Lake along the base of the Sierra Nevadas and down to the Los Angeles plain consumed twenty-two days. Panamint's rise gave his big wagons something profitable to carry on their way back from the seaboard.

From other directions the materials of civilization were also moving on the new camp. On the west side of the Sierras the Visalia & Inyo Road Company was organized, cheerfully disregardful of the 12,000 and 14,000-foot summits bristling directly in its track—the highest continuous ridgepole in the country. A closer inspection of those substantial upshoots caused that project's abandonment. But out of Pioche, Eureka, Austin, they were rolling: Panamint-bound trains of sixteen, twenty and twenty-four mules, each dragging four to seven high-loaded wagons, with a shouting skinner guiding each whacking procession by a jerkline and a well-directed shower of rocks.

"The goal was that winding, narrow-throated
canyon."

"In this wide street once swung fourteen- and twenty-four mule teams."

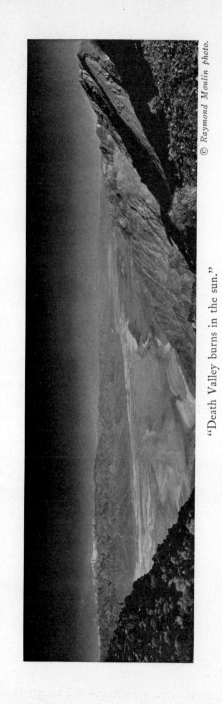

© *Raymond Moulin photo.*

"Death Valley burns in the sun."

Mining camp freighting as it was conducted in the Jingling Decades is still one of the unsung epics.

The mules, in the prosperous days of this trade, were sleek and gallant. The teamsters were hard-working, manly fellows of astonishing skill and a vocabulary that would light a fire. The big wagons developed for western service were virtual mountain ships, with the sides rising seven or eight feet, head and tail ends sloping at an angle of forty-five degrees, and the whole some twenty feet in length. This commerce reached its climax in the early supplying of the Comstock, when wagons stacked with tons of merchandise to a height of ten or fifteen feet moved in incessant parade each behind its dozen mules at a steady gait of three or four miles an hour, to the music of bells and the regularly timed percussion of the skinner's whip.

Such mammoth teams and wagons, choking the Sierra roads with moving piles of supplies, filling the uplands and hollows with their screech of brake-blocks, the imprecations of their drivers and the genial sound of their bells, had departed for the byways with the advent of the railroads. But in faded state they still served farther-lying camps, and were a prodigious factor in the far-flung scene.

Though in heavy going, such as lay toward Panamint, teamsters could no longer count on pulling twelve hundred pounds to the span, or reeling off twenty miles a day.

Yonder were waiting no teamster's caravansaries with blacksmith shops, shelter and provender at measured intervals. Out yonder there was not even water for the animals. And the goal was that straight-up, winding, narrow-throated canyon where there was scarcely room to pass a horseman. The little bells soon hung dust-caked and toneless as the Panamint-bound teams moved variously out across the San

Bernardinos, the Cosos, the Grapevines, and the Funeral Mountains. Getting goods to the Washoes, for all the trials of the road, had offered nothing like this.

Outfits pulled by bullocks were also to be noted. Though his days were numbered, the bullwhacker was still a vivid personality, and Clem Ogg as he swung twelve yoke of oxen into the Panamint parade, with the tongue of a second wagon run under the first and securely lashed; with ax and rifle hung on the side, and red wagon-blankets tossed atop the freight-pile, was typical of his breed. Clem was a big man with long and unkempt hair and his face was provided with a beard of cactus-like stiffness. His strong fingers gripped a whip with a three-foot handle wielding a lash not less than twenty feet long.

This lash was the scepter and its wielding the pride of Clem Ogg's vanishing order. From the handle its braided rawhide gradually swelled out to a girth, six feet or so from the starting point, about as big as a man's forearm. The remainder of its imposing length tapered until, a foot from the end, it became a ribbon-like thong.

This final foot was known as the "persuader." So diabolically accurate with the bull whip were Clem and his kind that an ox, at almost an astronomical distance from the driver, would double up as if seared by a red-hot iron when he heard its hiss.

Story is told of one of Clem's race that he bet a comrade a pint of whisky he could cut the seat out of a third party's pantaloons by whiplash without touching the man within. Faith in a bullwhacker's skill being what it was, the third party hesitated not at all to help along the bet. He put himself into position. The sinuous rush of rawhide through the air, terminating in a blow like a thunderclap, was delivered at a full twenty-five feet with great care and earnestness, but there ensued, it is said, the tallest jump ever put on record. The leaper with a yell sprang forth minus both hind pockets, but also minus much of the skin off the back of his lap, and the Clem Ogg who had over-calculated by just a leetle mite was heard to murmur sorrowfully:

"Hell! I've lost the whisky."

It was Clem, with Al Workman and Dan Paine, who got together for the Panamint trek the hugest outfit of all. Fourteen colossal wains of the pre-railroad days of the Comstock were drawn out of their retirement, divided into three sections each with its procession of mules or oxen; and the whole under its trio of captains set out from Wadsworth, Nevada.

At Lida the outfit took on 45,000 feet of lumber. Without road or map the vehicles then pushed on for the new silverado, the sand clouds of that jornada lying over the waste like white tornadoes. Up the abrupt, curling canyon the three sections finally came, beasts straining and drivers hurling blacksnakes and vocal suggestions; over the lip of Surprise Valley and into the hanging gully of a street.

"The rush to Panamint is increasing in volume," said San Bernardino's *Guardian* on November 12th, and "Twenty-four carpenters left San Bernardino on Monday for Panamint."

A week later, speaking of San Bernardino, "Business of all kinds is carried on at a gallop. The stores are crowded; hotels full; mechanics overworked; teamsters all employed; laborers at work; and everyone jubilant."

At Bakersfield, editor Joe Acklin pointed with derision to San Bernardino's "roundabout way to reach Panamint." He prophesied darkly that "their teamsters will recognize the fact about the time they are called upon to slaughter their first mule in Death Valley to sustain life while their wagons and machinery are settling out of sight in the alkali beds."

Despite this sweat and lather of onward movement, timber and machinery were not moving in fast enough to suit the galvanic Stewart. The emergency brought him whirling down to San Bernardino in mid-November, an animated dust cloud, having made the 166-mile trip from Panamint on horseback in forty-eight hours with a single change of mount.

He found the gateway town clogged with Panamint-bound freight and Meyerstein had a hard afternoon of it.

"The distinguished gentleman is the very personification of energy and endurance," admired the *Guardian*.

# THE COMING OF THE STAGECOACHES

THE road from Los Angeles to San Bernardino was white under August starlight, fenceless and deep in dust. Tom Peters was on the box of the stage, a kerchief knotted over his nose and mouth. The curtains of the vehicle were down to keep out some of the choking powder. Seven passengers rode with Tom Peters, all of them inside. Three women and two children, San Bernardino-bound. Two males on their way to Panamint who would travel with Tom as far as San Bernardino, and the rest of the way by buckboard.

The coach spun along, as silently as though on velvet. The four horses were fresh. Overhead the diminishing comet streamed its way.

A talc-laden bush at the roadside shook curiously and divided into two halves. One of these halves took the outline of a man, stepped directly out into the road and bellowed "Halt!"

"Hi!" yelled Peters as he sensed the fellow's meaning, and raised high his fourteen-foot whip.

"Stop them horses!" The alkali-coated shape made a clutch for the off-leader's head.

Tom Peters had sometimes wondered what he would do if he encountered genuine western road-agentry. Such a thing had never happened on his run. Now he discovered exactly what he would do. The buckskin whip uncoiled, then banged like a ten-gauge loaded with five fingers of buckshot.

"Hey!" The interruptor had spilled head over heels, the coachwheels nearly creasing him. He sat in the road and popped his revolver at the flying vehicle.

"Up, Nellie! Giddup, Gray!" Tom Peters had dropped to cover beneath his seat, and was guiding the team on one knee. The outfit pitched and swayed. Suddenly Peters' world split. A second depredator had appeared from the other side of the wide road. This one sent a shot into one of his horses, which screamed and stumbled, regained footing and struggled on.

Some hundreds of yards the coach and four continued. Then the stricken horse crashed. Peters leaped down and cut it from the traces. Thereupon he climbed back, gave a challenging yell and proceeded onward with three steeds at a furious gallop, the bewildered passengers doing ground and lofty circus work inside. They were extricated at San Bernardino, after some argument as to which arms and legs were whose. Half the town turned out for an exciting man-hunt. The road-agents were tracked to a little cabin in Cajón Pass and there they were collared, after laying low a deputy sheriff in the rumpus.

Action on the routes to Panamint had begun.

Direct stages for Panamint got into swing in the same month, a weekly vehicle going on the run from Indian Wells. The service was stepped up to tri-weekly in October, though a private lettter published in the Oakland *Transcript* warned travelers: "The road from Bakersfield to the diggings is about as rough as any in all California. The prices of '49 are demanded for all the necessities of subsistence."

On November 3rd the first stage from Los Angeles pulled

out for Panamint, crowded to the roof—fare $35, distance 225 miles, scheduled running time three days. It rocked up Soledad Pass, rolled across the Mohave plain, and pierced the El Paso Mountains through a weird gash where the red cliffsides were eroded into the shapes of organs, cathedrals, miterd bishops and congregations forever bowed in ossified prayer. As far as the Indian Wells turnoff this was the well-grooved route of Nadeau's Cerro Gordo bullion teams. Waterholes known by such names as Mud Springs, Barrel Springs and Cow Holes relieved it at fifteen or twenty mile intervals, their brush and adobe stations providing rough meals. As for beds, "Take your pick o' the sagebrush, stranger. Our rooms are never filled." Then, with the morrow, on and on, where endless parallel rows of little cardboard mountains raised their many-colored summits into a blue-hard sky. The troughs between the mountains were an interminable succession of bouldery slopes, silty bottoms, and playas that were now firm and smooth as skating rinks, and would be marshes of epsom and glauber salts and soapy tubs of soda when the rains commenced falling.

San Bernardino got its first Panamint stage away on November 15th, with eight passengers aboard. Shortly thereafter, Rubottom's hotel at Spadra on the road out from Los Angeles was reported "so crowded with guests that it is not unusual for persons to sit up until the stage goes out at 2 o'clock in order to get the beds the passengers have vacated."

Up the steep Cajón Pass to Heber Hunnington's station at the first crossing of the usually dry Mohave River, where Mrs. Hunnington's cooking bore a reputation; then to a second crossing where Barstow now stands; west and north to Black's ranch at the tip of a dry lake; on to a good well four miles farther, where the travel-wise quaffed deep, and from there by heavy road up a dry wash—no more water now for twenty-six glaring miles—with Pilot Knob pushing up the horizon; the Knob hour after hour looming ahead, ever drawing the wayfarer on, never seeming to lose its distance —"Seems like we're holding our own with it, anyhow"; the Knob's footings won at last, and for reward a fine well sunk deep in its shadowed granite; thence through a sun-baked gap with the Slate Range on the left and the Pana-mints on the right, and Panamint Valley's sterile sink lay beyond, with fifty-five miles and eleven hours yet to go before Surprise Canyon would be won—and water available along the way in just two places. Such was the "road" from San Bernardino.

Beyond Hunnington's the cookery also had its drawbacks, being usually performed by the driver. How many of the fat native lizards known as chuckawallas went into the broth is not of record, though "mock turtle" is a name that can cover a lot of fauna. But when Jack Lloyd told his hungry passengers that it wasn't the season for mock turtle and pro-posed "desert turkey" in its stead, cookery under the grease-wood reached a new high.

It was a recipe, Jack said, which he had learned from a prospector, since deceased. Not until later did the idea occur to his dinner guests that perhaps it was the recipe that had pole-axed the prospector. Jack disappeared while firewood was being gathered, and was heard banging his gun; re-turned with a fowl duly beheaded and skinned, and grilled it after a method of his own. It was only when his passen-gers became painfully distressed in and on top of the rocking coach, and Jack was observed reaching for his gun to fell another circling beast of the air, that the evil truth dawned . . .

"Buzzards ain't near as bad," protested Jack Lloyd, "as canvasback ducks that's been livin' around Big Lake over yonder. After three-four days o' that water they taste so funny that a coyote after taking one bite goes out and buries his teeth."

The stages were of all kinds: buckboards, "jerkies"— wagons with hard boards laid across for seats, and baggage stowed behind—double-decked Troy and Concord coaches, and strap-cradled "mud wagons."

The Troys and Concords, which by all odds were the favored conveyances when there was any choice, were relics of the vast Wells Fargo stagecoaching empire of a few years before. They had room for nine passengers on top, where plenty of sunshine was guaranteed, and nine inside.

Well-worn hair cushions on broken springs were reminiscent of what had once been comfort. Bullet holes in the landscaped door-panels told of past adventures. Multiple-strap "thorough-braces" between body and running-gear still absorbed some of the jolts. Dipping into arroyos, climbing up and out; pulling up after endless hours at some water-hole represented by a spring under desert willows, or a bucket attached to a fifty-foot rope—with a change of horses led out from a cactus-brush corral—and on again, and still on, they tumbled and heaved.

Night hours were especially favored for traveling, for then the air was almost always delightfully cool; though at times the driver would pull up, point to a curious curtain of little spangles bobbing and glancing in the starlight, and recommend a cautious tucking-up with blankets, especially over the mouth and ears. For this was the dread "pogonip" or frozen mist, to which Indians and desert men ascribed all bodily ills, especially pneumonia.

There were other travelers' hardships besides heat, chill, hard fare, glare, monotony and the pogonip. The Bakersfield route, ascending the Sierra, offered a succession of sharp turns and tremendous declivities. While Tom Peters was distinguishing himself to southward by outrunning bandits, a contemporary on the northern route tried to cast his coach and six too finely around a string of freighting mules. His eight passengers, seeing the narrowness of the road and the nicety of the job ahead, had accepted the driver's invitation to get out and walk. Stage, horses, and reinsman all rolled, tumbled, somersaulted down a steep bank a hundred feet or more: driver seriously injured, everything wrecked.

Three days later the Sierra stage again took a header. Though it was full of humanity this time, the spot was better chosen; there were many bruises but no serious damage. But on the night of October 20th, with E. P. Raines, R. C. Jacobs, D. P. Tipton—one of the Panamint originals—and John Searles of Borax Lake among the complement of the very full vehicle, a fore wheel struck a rock during a smart downhill trot and the pole snapped. Over and over went the Concord on the downhill side in a clatter of ash, iron and leather. This time the casualty list included a broken shoulderblade, a couple of broken legs and two or three broken heads, including the driver's. Two nights after, an ascending stage was caught and marooned for hours in a gale which lashed the southern Sierras and raised Kern River two feet.

Foremost at Bakersfield to benefit by the rising tide of Panamint travel was Monsieur E. Escalet of the French Hotel. His white-painted inn was now more than ever the stagecoach center of that end of inland California. With bustle and shine the leather-hung vehicles departed his door for points north, east and south. Back to his two-story, square-pillared veranda they returned, stages and horses on the keen run, brakes screaming, and passengers and outfit all one general dun hue, the overlay of the pulverized deserts.

M. Escalet loved this rush and bustle, now so happily doubled. He increased the size of his inn and he increased it again. Still his tables and beds all continued occupied. It should have caused him much contentment. But despite this liveliness and custom M. Escalet was not content. He toured his kitchen in a state of absent-mindedness and made regrettable mistakes with the sauces.

At last M. Escalet could stand what was ailing him no longer. He pulled off his apron, hung up his chef's hat and announced that the hotel with all its furnishings, viands, liquors, cigars, and good will was "For sale at a very low figure, the present proprietor having decided to try his fortune in the new mining district at Panamint." Presently M. Escalet, too, was one of the dun-hued passengers rolling off beyond the high blue mountains.

Joe Harris had a cozy saloon business at Independence. As Panamint-bound traffic down Owens Valley increased, Joe thanked the new camp and did rather well for himself.

Down from the north at about this time, with Panamint his ultimate goal, there came moving a bushy-whiskered tramp with a traveling technique all his own. On arriving at a stage station he would call up every lounger and fellow passenger and order drinks or cigars for all. After the usual ceremony of moustache-wiping, he would wink at the barkeeper and say "That's on me."

The astonished cocktail-diluter would inquire, "Who in thunder are you?"

The hobo would then lay one hand on his revolver and

reply heavily—"My name is Poker Bill; I have traveled all the way from Omaha on this." And from Omaha west it appeared to have been a formula as good as cash.

Such was the word brought by the stagedrivers to Independence in advance; and a few days later the man from Omaha came along. Joe Harris set up refreshments and the regular performance was enacted until Poker Bill started to explain his mode of travel.

Thereupon he found himself looking down the muzzle of about the biggest hand-gun old Sam Colt ever manufactured, and heard its proprietor informing him that he had routed himself in that peculiar way far enough. Poker Bill perceived that his ticket had run out and paid his bill like a man, remarking feelingly however about people being so particular over trifles.

Joe noticed that he still had a roll left. With the Colt in view he suggested that everybody might be receptive to another round. Again Poker Bill hastily agreed, asking only that he be allowed enough cash to get himself the rest of the way to Panamint.

But that was about Joe's last transaction at the county seat. In his veins, too, the Panamint yeast was working. He placed some bottles on the bar with a cigar box to receive the money, stuck up the sign "Let your conscience be your guide," put the rest of his stock into a wagon and hurried out of Independence on the trail already taken by most of his townsfolk.

Aided by subcriptions from fellow-Inyoites, John Shepherd of Owens Valley set to building a new toll road around the south end of the big lake and over the Cosos, bringing Panamint several hours closer to Lone Pine and Independence and affording a short cut for the travel now moving toward the new camp from Benton, Aurora and even distant Carson and Virginia City. With October's end, J. G. Dodge of Lone Pine set out to start a line of stages over this new six-station road, promising his passengers a one-day journey for the $25 fare.

But on his first trip Dodge became helplessly stuck in sand on the steep sides of the Cosos, was pulled out of that by his passengers and sank again to the hubs in a potash marsh. Next day he returned to Lone Pine to pronounce the defenses of the Panamints still impregnable from that aspect. Shepherd engaged every able-bodied man he could find and attacked his road anew. The next time Dodge pitched a stagecoach at it he rode over all obstructions and up to Panamint in triumph.

In October one T. S. Harris, no relation to Joe Harris of Independence but a printer with experience on a Sacramento newspaper and a short career behind him as minute clerk of the California legislature, boarded the stagecoach at Lone Pine. Into the rear boot went heavy impedimenta. With a concertina in his arms, Harris climbed up beside Driver Jack Lloyd. That distinguished buzzard-hunter seldom held one job long and now had shifted over from the San Bernardino route.

"I hear," said Mr. Harris, who was sociable, "you've got quite a camp up yonder."

The vehicle jolted along. Driver Lloyd, with the dignity of his whip-wielding order to maintain, was inclined to ignore talkative passengers for the first twenty miles or so. Harris waited hopefully, then added:

"In Panamint."

Onward rattled the stage, and silent remained Jack Lloyd. Over a couple of rises, and down the slopes beyond. Then the driver must have concluded to relent, for he made certain facial rearrangements and responded:

"Yup."

It was progress, anyway. Journalist Harris, having got one syllable out of Jack, hoped for more—perhaps a flood of three or four. He proceeded briskly:

"What every town needs is a good, live newspaper."

Two more hills and a couple of herbless arroyos passed under the iron-shod wheels. Then Lloyd looked afar over the landscape and answered:

"Mebbe."

This thing was developing into a dialogue. It might even grow into an argument. Harris persisted:

"I'm going up there to start one."

Not more than a couple of ranges and a valley intervened that time before the driver queried:

"So-o-o?"

"Yes," expanded Harris. "Four pages to begin with—later on, eight. It's to be called the Panamint *News*. I've got the type and press back there in the boot. It'll be published three times a week."

Day lengthened, and the shadows of evening went from violet to cobalt, before Driver Lloyd finished ruminating all that. What he finally said was:

"H'mp."

At noon the next day, when the Argus Range was being surmounted, he halted his steeds and pointed to a rocky mound with his whip.

"Thar," he said.

"Thar what?" asked Harris, with the journalist's eagerness to master the lore of the country.

"Whar we buried him. The first Panamint editor."

Harris eyed the pile in surprise. He had not heard of a predecessor in the locality. He would like to hear a little more. What had carried the gentleman of the press off? Driver Lloyd almost grew glib.

"He war shot."

Evidently annoyed at his own loquacity, the jehu then paid complete attention to his horses. Harris cleared his throat several times, but Jack's face remained stolid. The stage was out on Panamint Valley when the driver again extended his whip, pointing to a clump of mesquite on the far side. Again he remarked:

"Thar."

"Well, what happened over thar?" The man of types and ink could scarcely contain himself.

"He war lynched. The second editor."

Suspicious, Harris eyed the mesquite clump and the driver. He decided this time in favor of the method of the countryside. He stayed totally silent. It produced results, for in not more than a quarter of an hour the commander of the equipage was pointing his whip again.

"That's where he laughed himself to death, I suppose— the third editor?" suggested Harris.

The rest of the ride saw the passenger playing festively on his concertina and Jack Lloyd helping the new editor in the singing.

On a previous evening Jack Lloyd's stage and six had rolled up into Panamint town with exceptionally frail cargo. Jack swept into town with a flourish, and sat his proudest, amid the acclaim which' had been his in new settlements before.

Martha Camp was a lady not inexperienced in the ways of new towns or the art of mining the miners. She occupied the seat of honor beside the beaming reinsman. Martha Camp, from the crest of her imperial finger-puff to the regal velvet bow that covered her instep, was aware of the importance of the post she had come to occupy. Glittering jet and spangles of steel beading, here narrowing to a rivulet, there swelling to a lake, flowed over her bust and down her back. It wound about her sleeves and flounces. It cascaded down her skirt and underskirt.

Cheers of the townsmen greeted this sumptuously up-holstered personage. Queenly was her acknowledging bow.

Not for Martha alone, however, was the salute. For she was attended. Behind and above the well-armored lady (how *would* all that jet and whalebone fare up here on the rim of Death Valley, come summer?) was a ring of handmaidens who were slimmer, younger and less Junoesque than Martha, but not one whit less at home.

Each was equipped with the weapons of her craft—an enameled pair of cheeks, an undeviating stare, ten inches of cotton-hosed calf and a flash of ruffled pantalettes. The skirts of these charmers were gathered in scalloped tiers high up at the rear. Their hair was carried well up at the back of their bobbing heads, whence it was allowed to descend in a tumble of ringlets. Forward of these chignons rode jaunty chip bonnets weighted with roses and birds; rode shepherdess shapes of black velvet with pink buds under the side-brims; rode fetching gipsy shapes—tarnished and battered no doubt on close inspection, but bravely borne.

With mincing steps the naughty baggages came down from the deck of Jack Lloyd's coach and teetered over the stones in high-heeled shoes of once-bronzed leather, fastened up the front by little buttoned straps. Bonnets tilting forward on their foreheads, bodies tilting forward at the waist, sterns built out by bustles and flounces to emphasize the Grecian bend, the saucy newcomers skittered through the throng like ambulatory camels. Exploratory slaps of their voguish southward aspects drew swift rebuke from little red tongues but disturbed neither smiles nor dimples, nor affected the spirit of good-fellowship which everybody bore for everybody.

Virginia City had its Virgin Alley. Bodie would make much account of its Virtue Street. Panamint was not to be outdone in chivalrous naming. The gay arrivals were convoyed by wide-striding Martha up a steep side ravine previously known as Little Chief Street, which the camp now gallantly rechristened Maiden Lane.

# 14

## THE COMING OF THE ANGEL

THEY are pouring in now by tens and by scores: miners, gamblers, h'isters, adventure-seekers, ladies to operate bawdy-houses and Chinamen to operate "washees."

Some are with money to invest in a mine or business. Some are fortune's favorites from the last boom camp, some are the overstayers who saw the last tunnels cave in and the last lights go out. Old neighbors meet with handshakes of greeting as though but a night had elapsed since they parted, perhaps years ago. Some have played in a run of luck and some have failed to turn up the cards. Hardship and adversity for most are familiar partners; but still they come, bringing to this newest commotion on its painted hilltops the same zeal and grand undying faith that have kept them plodding the trail from hubbub to hubbub.

After months of adversity, in which each new day has meant a roll from dusty blankets, a brew of strong coffee and a sputter of salt pork and flour-and-water flapjacks in the frypan,—a diet to be persisted in three times a day, until wealth or better times shall ensue,—no wonder whisky tastes good, and friends and tales seem fair, and a man clutches with gusto for a "Brigham Young cocktail": one sip and you're a confirmed polygamist.

Main Street's rude pitch is a jangle of braying mules,

brawling oxen, swearing skinners, high-piled wagons, high-heeled riding men, bullies, claim-jumpers, thimble-riggers, San Francisco brokers in "top-mast stuns'l collars," staring Panamint braves, and eager newcomers.

"My good man, will you kindly tell me the quickest way to get underground?" "Just step right in there, stranger, and offer to lick any man in the place."

"Quick, bartender, quick! Gimme 'nother! Hurry, man, something terrible's goin' to happen!" "Where's the money for the three drinks you've already had?" "Ha-ha! Haven't got any money!" "Well, if I don't fix you, you cheating son-uva——." "That's just it, barkeeper. Didn't I tell you somethin' terrible was goin' to happen?"

"Git your likker in there, pardner—they serve it with

"They are pouring in now." (Los Angeles *Express* and Inyo *Independent*, 1874.)

a whisk broom." "What's the whisk broom for?" "So's you kin brush yourself off after you come to and git up."

"Notis! Any man or woman whose animal trespasses on this here claim, if I ketches the same will have his or her tail cut off, as the case may be."

Yells a would-be bad man: "I'm a cut-off shotgun loaded with slugs. Who wants to see me turn loose?" As he picks himself up and reaches for his hat, "Here I am again! Guess I musta went off half cocked."

Belligerent drunks draw "deadlines" in the barroom corners, each claiming to be a "chief." It is presumed to be dangerous to cross the deadlines. A dark, square-shouldered customer steps unconcernedly across one imaginary toe-mark, turns a cold level stare to the bully who squawks about it, is promptly and cordially permitted to enjoy his trespass. "That's Jim Bruce. Used to be one of Quantrell's gorillers. He's death on deadlines. Shook up a 'chief' twice his size down at Hamilton by the slack of his pants and scruff of his neck like he was churnin' butter. Killed a man in Treasure City for the same reason."

A newcomer on a tall, fine horse gets as far up Main Street as Oscar Muller's Arcade when the desire for fellowship possesses him and he enters. Being uncertain of the caliber of his new fellow-townsmen, or perhaps being very certain, he keeps tight hold of the tie-rope which runs out through the door and around the neck of his mount. Mischief-makers spy this situation and, after debating whether to substitute a runtish jack or the porch pillar, arrive at a third and gayer decision and lead away the steed. When the celebrator comes out, still clutching his end of the rope, he discovers the other end of it around the stout waist of a bland and smiling Panamint squaw, already assuaged by a red half-pint and a white half-dollar for whatever may be her lot.

Still they come, and the trails down on the desert are white with their dust, the waterholes aglow with their bivouacs. "So you're goin' to Panamint too. How long have you been prospectin' this sagebrush country, mister?" "Fifteen years, stranger, fourteen years and ten months o' which has been spent by me just a-chasin' these bastard jacks." —"Pardner, if you're bound for Panamint why don't you pick your kit up off the ground and put in on your mule?" "Stranger, I take it you mean kindly, but I can't ketch my mule." "And whyfore, pardner, can't you ketch up your mule?" "Because that mule, dad-blister him and me and you, is just too indegoddampendent, that's why!"

"Madame Moustache," Elenore Dumont, arrives from Carson City to pay an inspection. She hasn't missed a boom camp in years, deals a cool game of twenty-one or faro, and looks as young as ever to men who have known her from Washoes to Bitter Roots.

The Rev. Mr. Orne of Owens Valley, concluding with accuracy that the new camp will be a rare vineyard for his labors, hitches up and drives over, taking two days for the journey. Conceiving also that the laboring will be long, he brings a substantial stock of victuals. Grieved is Mr. Orne when a scamp steals from his wagon a sack of cabbages and two sacks of carrots and leaves a note requesting him to pray hard for the soul of the thief. . . . The Rev. Orne holds services in Dave Neagle's tent and learns to be unconcerned when his "Amen" is echoed by a worshiper yelling "Keno!"

Neagle's resort goes in for varied entertainment. Last night it was the " 'Frisco Dance & Vaudeville combination"— songs, farces, drama, banjo solos and Dutch specialties, very broad. To-morrow night it will be a concert. A throaty soprano, large and blond, gives "Come Where My Love Lies Dreaming," varied with "I Won Her Heart at Billiards," and "I Hope I Don't Intrude"—the last quite kittenish. The curtain is pulled up by accident and a joyful moment is created when the lady is discovered changing her costume. " 'S'all right, boys—don't be embarrassed!"—the buxom entertainer, it appears, is a man in tights, stays, padding and flaxen wig. A disgusted customer drops the curtain with a rush by cutting its ropes.

Nick Perasich and a couple of friends open a French Restaurant, paying $200 for a tent and two lots. Their establishment does not go in for style, but as the proprietors do their own serving and each wears a gun in a visible holster strapped well down on the leg, there are few complaints from customers. Writes one newcomer: "My first meal in the town was supper, when I joined the rush of miners for the dining room. The man who brought in my mess was not sober, and spilled soup all over my clothing. I told him not to mind as I had another suit."

Neagle has lumber on the ground for his new edifice across the street. He has bought up the block of lots on Main Street adjoining Maiden Lane, perceiving that here will be the civic center, and is busy making money with his subdivision. Harris & Rhine, his neighbors, are preparing to receive a large mercantile stock when their bull team gets through the sand and up the grade. (Three Harrises have so far worked into this narrative. They are not the same.)

A few letters go south by the galloping pony. "If you can whack a sixteen bull team, hit a drill, engineer a wheelbarrow, deal faro or shoot, come right along. Otherwise, stay where you are." "You can find any game you want here and all claim to be strictly on the level, but the bottom of

Main Street is 1000 feet lower than the top and I haven't found a square foot that's level yet." "The weather is all right so far, but the camp's in a gulch. If a snow or rockslide comes or a waterspout hits the hilltops, God help our town!" T. F. A. Connolley, a pioneer prospector and resort keeper, writes down to the paper at Independence: "The future is bright indeed. It is our opinion that such an opening for the development of rich mines has never been known in the country before, and that the excitements at Virginia City and White Pine were as naught to what there will be here in the Spring."

The glowing glamour of Panamint by now is casting long beams across the sand. They reach southern Arizona and penetrate Tucson's jail, where four murderers are languishing. Promptly these heed the call of the rainbow trail and over the wall they go. Panamint lies five hundred miles away by the shortest line, but the camp's growing population is augmented a few weeks later by the safe arrival of the Tucson killers.

These and a hundred other walkers of untranquil ways hail each other in Panamint with the easy camaraderie of men who have shot and dodged and hated each other so long that they meet as old friends. Pioche and Eureka and Virginia City watch them go with relief outweighing all regrets; there is no record of testimonial dinners to those departing.

J. P. JONES did not leave the Atlantic seaboard for the scene of his new venture until August, being detained by an extraordinary number of persons who wished to advance the moneyed solon's fortunes by selling him something. Their proposals ranged from New York's fashionable St. James

Hotel to the patent for a new-fangled ice-making machine, both of which he purchased; and there were offers of farms, stables, ranches, mines and newspapers which he promised to investigate. He was finally greeted at the Nevada capital late in the month with bonfires and oratory and "those ostentations of joy," as the politically hostile Virginia *Chronicle* described them, "which the possession of an exaggerated amount of bullion will generally secure."

Fairly burgeoning with interests, which included a budding affair of the heart at San Francisco, Jones passed the concerns of Panamint over, for the time being, to Brother Harry and the able Bill Stewart.

People kept thronging about the great man wherever he appeared in public. E. P. Raines was usually in the offing. It annoyed him, no doubt, to see other people getting a hand into the Jonesian pockets, but at least he had had one scoop.

Meanwhile the open-handed J. P. acquired a mine in Arizona, other mines in eastern Oregon, a steamy and very Mohammedan Hammam bath on San Francisco's Dupont Street, and 120,000 acres of salt marshes on the north shore of San Francisco Bay. For this last he ordered $200,000 worth of dikes—envisioning a little Holland of 120-acre farms. Colonel Baker and General Beale finally plucked him southward to inspect a proposed dominion they were holding for him on Santa Monica Bay.

Raines would have enjoyed being one of this party, which coasted south in the revenue cutter "Oliver Wolcott" and was taken from Wilmington to Los Angeles in a special train. But the Jones sense of humor directed otherwise, and this particular "Colonel" to his own disgust was left to follow on the regular steamer as major domo to a chattering batch of Chinese road-makers.

Senator Stewart met and conferred with Jones at Los Angeles. The moneyed junior Senator and senior partner approved of all that had been done at Panamint; he had received charming Georgina Sullivan's shy "yes" just before leaving San Francisco and was in a mood to approve everything.

Following his conference with Bill Stewart the happy man left for Shoo Fly Landing, where Baker, Beale, Temple and young engineer James V. Crawford wanted him to behold the water terminus of the proposed Los Angeles & Independence Railroad. It would be just the place for shipping the ore that would soon come pouring down from Panamint like a torrent leaping for the sea. . . .

Jones agreed heartily, and subscribed an initial $200,000 for the railroad. Purchase of a substantial share in the San Vicente y Santa Monica rancho had already been confirmed. He hurried back for San Francisco and pretty Georgina on the "Wolcott" that evening. Stewart had returned to Panamint.

On October 25th Senator Jones passed through Bakersfield on his first personal visit to Surprise Valley, and found the rush in full flood. Scattering benefactions as he went, the genial man paused long enough between stages to acquire, for $150,000, controlling interest in the Sumner gold mine at Kernville on the mountain-tops. It turned out subsequently to be one of his big money-makers.

Beyond Indian Wells, where the granite mountains flattened out into a sea of deserts, he settled himself into the hard rear cushions of one of Panamint's sand-liners. The stage was crowded. From the Wells the way led out across a greasewood expanse friendly under skies of Indian summer.

Salt Springs, the first stop for water, was now decorated

with a cabin as befitted a busy stage station. Beyond that white-crusted ravine with its lethal seepage were twenty-five miles of rough going. From sunbaked passes he moved on down to the white "lake" where the industrious Searles brothers were tending their vats and where drinking water, hauled several miles by wagon, was free to all who came that way. The rising tide of Panamint traffic had brought to their door the immense gift of a wagon road, and in gratitude the Searles boys were even putting up frame shelters for the Panamint-bound—everything cordially gratis. They were also expanding their borax operations and now managed sixteen sagebrush-fired vats. It is remarkable that Senator Jones in the presence of all this energy and optimism did not buy the whole lake and reduction works in passing.

The borax sink was gradually left below and behind. Eight or ten hours of more lurching and buffeting, of dropping heavily into chuckholes and jolting out, and the illustrious man was setting foot in the town to which he had given hope and nourishment.

The men of Panamint were eager to see this member of their order who had become one of the most talked-about personages in the land. The reception committee included more than one rough admirer who had known J. P. Jones between fortunes in half a dozen diggin's.

These now perceived descending from the stage a ruddy-faced deus ex machina overflowing with health and energy, fitting associate to the bulky Bill Stewart who greeted him. The muscular torso was grown a trifle rounded, the large head and brown beard slightly gray. But the memory was acute, the grip powerful, the dark eyes kindly and shrewd. "Howdy, Charley! Still at work on your first million, or have you set it aside awhile till you clean up your second?

Well, Ed Claiburne! How's the game left leg? So you've gone in for a wooden one—cuts down rheumatism by half, eh? Don't slip off, Andy Clummer, around the corner of the house like that. You owe me twenty dollars, you scoundrel, and I'm going to lend you twenty more to pay it with, and then we'll all go in and watch you buy—any objections, boys?"

Among those who greeted this most popular man in America was one who had been sheriff of Trinity when J. P. was his deputy.

As the two old comrades gripped fists their minds may well have swept back to a day, twenty years before, when a dozen white men had stood off, back to back, the attacks of the Klamath Indians for sixteen hours. J. P. Jones and Captain Messec had acquitted themselves with fortitude through that long, hot day—indeed, there had been little else to do about it.

Facing his old chief, whom he had succeeded later as sheriff,—what years, what miles, what ups and downs of fortune had since intervened!—Jones now appointed the scarred battler his superintendent of the present operations.

The bearded angel found that his settlement was one of no great beauty or many houses. Most of its 700 inhabitants were still living in tents or camping on the ground. Cabins were continuing to rise by the formula of piling up the loose stones into little walls. It was easier to do that than to continue sleeping on them. Lumber was coming in by mule or bull teams that dined, on arriving, on $400-a-ton hay sold by the sack. The air was electric. Real estate in that region of vast vacancy, at $500 to $1000 a lot, was changing hands almost as fast as jackpots, and jackpots were moving.

With both chief sponsors on the ground, things proceeded to hum with redoubled vigor. Bart McGee's toll road up the canyon was bought by the two senators for $30,000. Many new claims were taken over. Jones was waited on by a deputation of mine workers who wanted stock in his Surprise Valley Mill and Water Company at bedrock prices, to be paid for half in cash and half in labor. Their proposal was accepted and three thousand shares were so distributed, with options on a like amount more. There was a spontaneous surge to name the camp "Jonestown," and though the name failed to stick, the affable visitor acknowledged the compliment in the proper frontier way.

"A gentleman just in from Panamint," chronicled a Los Angeles dispatch to the all-observing Sacramento *Union,* "reports the district in a fever of excitement. The mines are developed sufficiently to warrant the belief that they will rival in richness the famous Virginia City mines. The town is filled with saloons, and gamblers are doing a splendid business. He says he saw $2,000 in one pot. A pretty good indication of good times—so old Californians say. New buildings are going up rapidly. The lead owned by Stewart, Jones & Co. is said to surpass, in richness and vastness, anything ever struck before west of the Rocky Mountains."

# 15

# THE GHOST OF LITTLE EMMA

THE mellow weeks of November, '74, before first snow flies, find Panamint in the full surge of joy-of-living. Tunnels are being driven into the cliffs and are bearing out predictions. Jones and Stewart's Surprise Valley Company still is buying; rare is the camp with the backing that this one has.

Main Street is alive with rumors. Interested knots discuss each new strike or deal. The Hudson River mine on the south slope has been sold to the Surprise Valley Company for $25,000; Pat Reddy is the lucky owner of 110 feet. Dave Neagle has just come down from the confluence of Woodpecker and Sour Dough ravines with samples from a mine discovered by a couple of stage-robbing rogues; the deposit rivals the top of original Ophir on the Comstock and will bear that immortal name. The big horseshoe of high hills above the camp requires four or five days of scrambling to cover thoroughly and there are prospects, stakes, monuments everywhere.

A bulletin board has been set up in front of Will Smith's notarial and justice office. It records the local news and the antics of the distant Comstock. Groups are continually forming around it, scanning the quotations of Crown Point, Belcher, Mexican, California, Con Virginia—though, as everybody knows, the Comstock will be a back number when this camp hits its stride. . . . An oracle takes his stand and the crowd drifts his way, to hear him expound the more local prospects of Wyoming, Hemlock, Alabama, Little Chief. . . .

Into this rapturous paean pierces a discordant note. Somebody pins to Will Smith's bulletin board a clipping torn from the San Francisco *Bulletin,* that self-appointed guardian of the unwary investor:

## Is It a New "Emma" Looming Up?

There are two kinds of mining discoveries. One is quiet, silent and earnest. The vein or district is explored and worked. Sometimes capitalists are quietly invited to take an interest. . . . The second kind of mining is the very opposite of this method in every particular. It proclaims itself from the housetop. It is embellished with the names of diplomats and Governors. A newspaper or two or more sing its praises in a persistent and unchanging symphony. The great Emma mine of Utah is an illustration of the latter kind of mining enterprise. It is a notorious fact, however, that the most promising shafts which it sunk were in the pockets of the credulous English purchasers. Its most successful tunnel was run, strange to say, in Lombard Street, London. Then there was, of an earlier date, the great Mariposa job. The papers of the time were full of it. Wells Fargo & Co. were laden with the treasure which purported to be its yield. But the real mining was done in Wall Street. The only well defined gold vein that was struck, was located in that financial thoroughfare.

It is not necessary to inform the world that this line of mining pays very well. It is an old Mexican maxim that to work a silver mine a gold mine is necessary. For this sort of silver mining nothing is requisite but brass. There is a district in the southern part of this State, called Panamint. . . .

William M. Stewart and Trenor W. Park were two of the promoters of Little Emma, the Utah mine which was named for one of Brigham Young's wives and had already produced several millions in white metal. According to current charges the vein was nearing exhaustion when it was bought up by silver-mad Britishers, on recommendation of a well-bribed financial editor of the London *Times* and of an equally tarnished American minister to the Court of St. James. The dividends that were promised ran to seventy and eighty per cent. But the mine, once transferred, abruptly pinched out; the last two monthly dividends were found to have been paid from borrowed funds; and the international tongue-wagging that followed had the Atlantic fairly whipped to a froth. Stewart and Park were at this moment being sued by the angry losers for $5,000,000.

> We shall rejoice if any really valuable discoveries should be made there [continued the *Bulletin*], but the agitation begins in the suspicious way. The shapes of some of the Emma and Mariposa speculators are flitting about in connection with Panamint. . . . The old operators and old apparatus are distinctly visible. . . .

The Mariposa deal as referred to had been another questionable affair, engineered by Park with the help of the national hero John C. Frémont.

Other journals dissented from the *Bulletin's* cold scorn. They pointed out that Jones, Park and Stewart were investing their own funds. A few days later the *Bulletin* relented, and the editorial clipping that fluttered from Will Smith's tent-front declared, under the heading "THE NEW ELDORADO," —"Panamint is the new Mecca for all the adventurous spirits of the Pacific Coast. The man who came to California in '49 under the mining excitement, and has continued his peregrinations up and down the coast ever since, with occasional wanderings into Utah and Colorado, has now set his face toward Panamint. The emigration from Nevada commenced months ago. The fever has reached as far as Idaho and Montana. . . . Some who have visited the new district are occasionally heard to say it is a second Comstock and will rival in its production the mines of Storey county, Nevada. These early informationists are, no doubt, excited enthusiasts, but such men are always the pioneers in discovering new mines. Once in a thousand times it happens that they prophesy correctly.

"Panamint is now undergoing all the experiences incident to the establishment of a mining camp. Although a population of seven hundred men and a few women are gathered there, only one comfortable dwelling yet exists. Town lots are being held at speculative prices, and we hear of squatters relying upon the virtues of a good shot-gun as their title to certain favored locations. . . ."

Did Jones and Stewart conscientiously believe they had a bonanza lurking behind those colorful cliffs? Perhaps so, perhaps not. Certain it is that they did not intend to deny to the public a share in their mines, particularly in those of least promise. No schooling on Mother Lode or Comstock could possibly have taught them any such idealism. If you can sell your treasure before you've mined it, and get double, triple, a thousand times its worth, why wait? You don't dig shafts, drifts, stopes and winzes with conscience and benevo-

lence. You dig them with capital and assessments. Ownership and risk must be divided with him who will go in. The
fascination of the game lies in the fact that no human being
can positively know what lies an inch below the last point
pick and shovel have explored. The defense of it lies in the
fact that without somebody taking chances, mines could not
be developed; and, anyway, while the uncertainty lasts
everybody has a great time. In the case of Panamint the
opportunity for stock-rigging was certainly present. It was
all very far away, difficult for the public to examine, held by
locators who through circumstances could be readily bought
out: offering a chance for a tight little monopoly such as the
Comstock never had been. Just the sort of a package to sell,
in particular, to the gullible British. If only there had not
been that uproar over the Emma! So Jones and Stewart
bought, and bought rapidly.

But whatever the private plans of Jones and Stewart, with
each new stage the rivals of the stock boards are arriving.
Three years ago they had a grand time of it, selling half
the rocks in Nevada while Jones and Hayward and Sharon
were running Crown Point and Belcher up to nearly $40,000
a foot. Hundreds of wildcat mines were carried up in
sympathy. Now, with circumstances on the Comstock shaping toward a terrific revival, they have been looking for
other prospects suitable for tailing the kite.

Since July they have been eyeing Panamint, and to eye is
to act. Four of the original locators, Copely, Parker, Tipton
and Gibbons, each valued their claims at $25,000 yesterday.
Today they strut Main Street as directors in very fancy-
sounding companies; the claims have been incorporated for
$24,000,000. Worries the Bakersfield *Southern Californian:*
"Quite a number of incorporated companies have located

their works in Panamint and their principal places of business on California Street in San Francisco. We hope this will not detract from the really good mines at Panamint."

Not all of the arrivals from California Street are vendors of manufactured glitter. George Hearst is one who looks in on the new camp. He does not mince about in Congress boots with elastic sides, in gray trousers or in tailed coat; his feet are durably shod and his baggy pants tucked in. What he sees he will report to his associates Tevis and Haggin. They are a firm who keep what they buy. Ranging over canyons and ridges, George Hearst does find what he is looking for, and the Christmas Gift mine changes hands.

Another who looks in is alert, quick-eyed Elias J. Baldwin, of straggly chin-beard and receding, brushed-back hair. No child of chance is Lucky Baldwin. He operates for himself alone, and little escapes his glance. Years ago he drove with horses, wagons, merchandise and passengers across the plains to the gold country; he made money on everything he touched while many men were starving; he purchased and ran up profits with the Pacific Temperance House on San Francisco's Pacific Street, a thoroughfare not exactly famous for temperance; as a sideline he made the bricks for the government's fortifications on the Golden Gate. In that deal his profits were as solid and as many-sided as his bricks, for he also boarded—at a rakeoff—his army of brick-making workmen. Later he plunged heavily but judiciously on the Comstock.

When not studying the veins of the Panamints, Lucky now sits and draws up, on scraps of paper, designs for the new hotel he plans to erect down in San Francisco in the spring. It will have a mansard roof, Corinthian columns, classic cornices, towers, more bay windows than there are warts on a lizard, and a theater. Also he plans a stock ranch in the south. Blooded horses for a pastime, but for a profit, mules. They're the one permanent necessity in this western country. Meanwhile he finds the Panamint claims all taken up, but there may be something worth looking into over yonder in those Cosos. . . .

On October 21st, up at Virginia City, Jim Fair takes the clamoring newspapermen at last under ground, and there it is —the Bonanza of Bonanzas, and *real*. If a prize so long invisible can prove to be so actual, what shall be said of far-away cliffs down on the edge of Death Valley that flaunt their silver to the sun in visible ledges hundreds of feet long and twenty feet thick?

On November 28th, California Street is hearing of the "Idaho Panamint Silver Mining Company," capital $5,000,-000. December 2nd produces the Maryland of Panamint, $3,000,000. On December 10th there is a perfect explosion of Panamint incorporations, seven in all for a total of $42,-000,000, and the Panamint Water and Toll Road Company for $200,000 additional. Nor are these the end. In February the "Panamint Consolidated Mining Company" will be making its $6,000,000 bow and in April the brokers will be marching down to the gravel ramp at the canyon's mouth and there staking out a 640-acre townsite for the fantastic "Panamint Mining and Concentration Works," capital stock $5,000,000, its chief claim to attention being that it is within a few miles (the prospectus does not say vertical) of Jones and Stewart's holdings, and "Books for subscription are now open at the offices of the company." In all, there are manufactured $86,000,000 in certificates (for which, of course, there will be little paid-up cash but great prospects for assessments) in behalf of a gorge in the rocks at the center of what recently was the most unwanted region on earth.

Jones and his entourage left Panamint toward October's end, their leave-taking accelerated by the tales coming down from the Comstock.

As the affluent Senator went down the hill, his coach in a narrow turnout passed the upbound vehicle. T. S. Harris, Jack Lloyd's outside passenger, was making that entrance with his concertina, a light tent, a few cases of type, a little Gordon job-press and a first-class supply of resolution. J. P. Jones halted his descending coach to beam benediction and promise some advertising. The camp up yonder by now had twenty-six frame buildings, many huts and stockades, six stores, eight hundred inhabitants, two bearded angels and several whore-mistresses, and it was high time such a promising camp also had a newspaper.

Stewart stayed on at Panamint a little longer, though a lame-duck session at Washington was calling. The senior Senator hated to leave this robust atmosphere. Stewart Castle might be the finest mansion in the land, and California Street on this opposite side of the continent a most exciting thoroughfare, but a cot out under stars had its own kind of comfort, and the clean wind of many deserts blowing up Surprise Canyon was breath of life to the man who at forty-seven was out to wrest his third fortune from the rocks.

Only one characteristic of a boom camp was lacking. When Bill Stewart made that August trip to San Francisco, he had paid a call at a stout, iron-shuttered building at No. 320 Sansome Street. But here he had met with less than his accustomed success. Bill Stewart was, in the course of his busy career, a frequent attorney for Wells, Fargo & Co. And Wells, Fargo & Co. was the treasure carrier for all the West. Few indeed were the places into which it did not send its messengers to fetch down bullion satchels, sacks and coffers. In Bodie, three years later, Wells Fargo would meet a situation similar to that offered at Panamint by putting three armed messengers on the treasure coaches—one on top, one beside the driver, and one inside. Other shipments would be accompanied by squads of as many as seven armed men. For escorting wealth from difficult camps down to places of safety the big express concern charged one per cent of the value of the cargo—and Stewart, thinking of his narrow-gated camp and some of its residents, was eager to pay the price.

But John J. Valentine, general superintendent of the express house, had heard of Panamint's setting and the style of some of its citizens. If reports from central Nevada were true, his stages had already had their fill of the latter. Despite Bill Stewart's eloquence, which Wells Fargo much respected and often employed, General Superintendent Valentine and his president, square-bearded Lloyd Tevis, remained unmoved. An agency up there, perhaps, if some local merchant cared to sell Wells Fargo's drafts and cash its paper for a commission. But express coaches to that suburb of hell, Mr. Stewart, and treasure chest down therefrom?

No.

The problem of getting the silver down from Surprise Valley, out through those grim portals and across the miles of sage remained Stewart's own.

# 16

# THE COMING OF THE FLESHPOTS

PANAMINT was rough. Its single bootmaker did a good business covering feet against the hard cobbles that paved its paths. Two hundred Chinese were put at work chipping trails and galleries along the cliffsides to reach the mines, and Jim Bruce went up and down Main Street soliciting contributions for smoothing that thoroughfare. But Panamint stayed rough. Snow began falling late in November. It came at first as a gentle powder. Publisher Harris, who was camping on the ground, shook it out of his type cases and got them under canvas cover; nobody else was discommoded.

Followed a spell of genial weather in which the camp moved around in its shirt-sleeves, and newcomers from the Death Valley side reported that hollow as comfortable as a lady's lap or a pisco jag. Then winter roared up over the horizon and fell on the Panamints with force. Surprise Valley went to sleep in gentle rain, woke in a blizzard.

Ended now were the days of camping under the stars. The homeless ones crowded the tavern floors by night, to be swept out with cigar butts and torn cards in the morning. Those with hovels developed a lively habit of stealing from each other's woodpiles. Before winter was over it would be recognized form to conceal a half stick of giant powder in the cordwood, and to await with vengeful pleasure for the moment when a neighbor's chimney would make a leap through its roof. Fires in rapid order consumed Archie MacDonald's tent saloon, Weber & Heneche's barber shop, Burkhart's stock of watches and jewelry and Waltz' cobblery. The camp had plenty of gin left, it was careless about its whiskers, and the hours and their divisions were of little significance; but that bootshop *was* important.

To the din of never-closing whisky mills, the sharp ugly voice of occasional pistol and the stamp and jingle of arriving jerkline teams were added the continual clatter of trail-blasting and mining along the mountainsides. Rocks frequently went soaring entirely across the canyon, to thump against the opposite walls and set small avalanches sliding. "There is rarely a moment during the day or night," wrote someone to the weekly at Independence, "when the reports or reverberations of the shots are not heard."

A description of the town at this stage was brought to Eureka by James Hayes, who expressed himself as favorably impressed with its future prospects but pronounced life there extremely difficult at the moment. Lumber and provisions were high, there was not a lodging house in the town, and every newcomer was obliged to furnish his own blankets and look for a shakedown on somebody's floor. Meals were "a dollar each, with an extra charge for everything except bread and mustard."

Publisher Harris got his newspaper going on Thanksgiving Day, issuing the initial number from a seven-by-nine foot tent. Its four pages, each about a foot square, offered one D. P. Carr, late of the Carson *Appeal*, as its editorial associate, and pledged its objectives "To furnish the people of Panamint with the latest news; to give to the 'outside world' accurate and truthful information regarding the mines and district; and to make money." The next issue announced severance of Carr's connection and the third pronounced him a dead beat, an absconder and a "mis-carr-iage," who had collected three months' advertisements and subscriptions in advance and gone down the canyon with one man's blankets and another man's $18 pair of boots.

Panamint was at the top of its boom that winter of '74. Six hundred claims had been pegged off. Lots on Main Street had gone up to $2500 and $3000. Jones & Stewart had laid out half a million for locations and roads, and eight of the leading prospects were being tunneled with gopher-like energy. The great Wyoming mine was a yawning window let into the high south wall. Three powerful mules, each hitched to an iron-shod sled, were continually dragging its output down to the settlement. There the rich ore was being stacked up in sacks, ten tons a day. The silver in it would all be profit, for its copper alone was enough to pay for its outward transport by wagon or packmules, its rehandling by rail down below the rim of the deserts, its portage by Pacific ship, isthmian rails and Atlantic ship, and its ultimate reduction in British smelters. Assayer Meyer, going about the scene and making 292 separate tests, announced a grand average of $400.70 to the ton.

Two hundred and fifty of the town's populace were finding employment in the tunnels, building tramways from the mines and erecting ore-breakers, feed yards and boarding houses. The rest found occupation as they chose. A Masonic lodge was formed and there was excitement over an election for recorder. In the latter ceremony 900 votes were cast, J. M. Murphy wresting the crown from Dave Neagle; "but as the voting was done early, late and often," noted a participant, "no correct idea of the population can be had from that vote."

Cold weather halted the pony express. The swishing pad of hoofs in desert sand slowed down with first snows in the passes. The miners at Panamint were an active lot, but letter-dispatching was not one of their passions. More than letters also failed to come up the canyon; there was a genuine shortage of foodstuffs, and Editor Harris sighed for $2.40 that he was compelled to lay on the line for two heads of cabbage. A customer complained to Jim Brown, clerk at Harris & Rhine's store, that the potatoes sold to him looked frozen.

"If you don't like 'em," advised Jim, "you can repudiate 'em."

The customer went away, thought that one over, experimented a bit and returned in smiles. "I'd biled 'em and peeled 'em and baked 'em and fried 'em," he confided, "and they were terrible; but when I tried 'em the way you mentioned, they were just fine."

Rough, too, stayed that corridor leading to the outside world. A writer in the *Alta*, describing the road in January from his safe distance of six hundred miles, assured his readers: "The canyon is traversed by an excellent road . . . smooth and regular, and well laid out. Coaches may go from the top of the canyon to the mouth at full trot, without trouble or discomfort to travelers."

This was boosters' talk, as was soon discovered by an incautious teamster who neglected to hitch some of his horses behind for hold-back strategy. With all six in front he started down the chute. His brakes began to smoke, wheels spun faster and faster, and the wagon became a juggernaut bearing down on the slipping, sliding steeds which finally hoisted tails and shot into a gallop. In an instant the whole outfit was a six-directioned projectile. Score: one dead lead-horse, one swing-span with legs broken, one driver flung against the canyon wall, and one vehicle in splinters; and Billy Killingly, the inveterate horseman, swept from saddle as he met the avalanching runaway and all but obliterated. When Billy Killingly came to, Dr. Wells was bending over him.

"What is it?" gasped Billy.

"Broken leg. Hold tight—I'm going to set it."

"Be sure to set it in the shape of a horse," groaned the riding man.

Lewis Smith and passengers started down in a two-horse wagon. The horses started to run. Smith was late getting his weight on the brake. A quarter mile of mad going with ever-increasing velocity, and all up-ended in a mighty arc. One of Nadeau's teams came by and carried the men to Borax Lake. Coming to, they saw its white shimmer and lay back convinced they had been hurled clear over the Slate Range.

Editor P. A. Chalfant of the Inyo *Independent* went over to pay the camp a December visit. When the last six miles became practically straight up, and sitting on the seat of the coach somewhat like lying on one's back, the writing man got ready to make a spring. The vehicle and its exhausted horses, he felt, must surely slip back. And at one spot, slip back it did, while the horses stopped for wind and the driver was shoving down his brake. Editor Chalfant wasted no time on farewells. He leaped.

The result was a beautiful professional pi in the midst of chains, stamping hoofs, double-trees and wagon-pole. With dignity he allowed himself to be extricated and with magnanimity he wrote: "The future of Panamint is exceedingly promising. It will afford a great basis for stock operations; in this it will hardly stand second to the Comstock itself. . . . It would seem to be an almost absolute certainty that the pay ore runs down to immense depths. There can be no question that these are true fissure veins."

In January an inspired informant in the San Francisco *Alta* gave Panamint a population of "perhaps 2500," of whom the chief proportion were "miners, the balance being made up of merchants, saloon-keepers, and the inevitable sports who display a large wealth on faro and monte tables for those who indulge in the gentle games of chance." A few days later the *Alta's* describer reported that he was back again in Panamint, ascending Surprise Canyon by "a splendid wagon road," whereupon, after such mild and pleasant faring,

### PANAMINT APPEARS IN VIEW

No lofty spires, or steeples; no towers, no fine public buildings meet the eye, but it rests nevertheless on a matchless piece of enterprise, a six-months-old town beautifully located, solidly built, and eminently respectable in the character of its structures. A large number of families have transferred their fortunes to Panamint, and have settled permanently. They have built some very fine, commodious and substantial stores, restaurants, lodging houses, saloons, stables, residences, etc. . . . The store windows display goods with great taste, the interiors of the buildings are well arranged, and on the whole a stroll through Panamint cannot fail to leave the impression that the town is a permanent one. . . . It is only necessary to look at the rapid growth of this little town in an isolated section of the country, in order to get an idea of the activity which prevails at the mines, of which it is the blossom.

Editor T. S. Harris seems the likely panegyrist. He has just returned from San Francisco with new press and

SAN FRANCIS

## PANAMINT.

Its Immense Growth and Future Prospects—The
Great Mines Shown to Contain
Untold Wealth.

What Enterprise, Energy, Confidence
and Capital Have Ac-
complished.

Panamint the Centre of a Great Mining
Region.

Recent Rich Discoveries.

The Brey Fogle Ledge Discovered by
Panamint Prospectors.

The Los Angeles and Independence Railroad.

INDUSTRIES WHICH IT HAS AL-
READY STIMULATED.

The Great Borax Lake—Pre-
parations to Work It.

$1,200,000 Expended on the Panamint Mines—
Enduring Character of the Work Done.
And Rich Results Anticipated.

[FROM THE SPECIAL CORRESPONDENT OF THE ALTA.]

PANAMINT, January 25th, 1875.

I arrived here four days ago, after an almost continuous drive of fifty-seven hours. The journey
presents all the variety of California scenery, from
the luxuriant valley clothed in its robe of green to
the "rugged edge" of the bare mountain, and even
the mountain summit. And I am not out of the
world, or if you should think so then I may say I
am in a new world, recently sprung into existence
by the magic influence of wealth, energy, perse-
verance and an unflinching confidence in the great
future of Panamint and the surrounding country.
But before proceeding to give you an idea of Panamint as it is, or looking at its future, I will first de-
scribe the trip, which cannot fail to be of interest to
your readers. I took the Central Pacific cars at Oakland at 4 P. M., and went to Lathrop, where I arrived
four hours later. Here I took the Southern
Pacific train and sleeping cars for Bakersfield, where I arrived at 7 A. M. next day.
We made close connection with the stage which

---

types. His lyric disdains to add that the town contains neither church, schoolhouse, hospital nor jail—nor never will; that its populace is saved from something close to starvation by the bold arrival of one Patten with a thousand bawling head of mutton, driven over snow and sand for a hundred miles in the teeth of a blizzard; or that Wells, Fargo & Co., that universal business agent of all the region from the Missouri River to the Pacific Ocean, still declines to have anything to do with the sinister settlement. Yes, Panamint is rough.

AND PANAMINT is tough. That is the natural working-out of the immurement of several hundred men on a mountain shelf with nothing for amusement save raw scenery and raw stimulants. The camp is set in a gash in harsh rock mountains. The mountains are set at the core of harsh concentric deserts. Men die on the trail to Panamint. Men die on the trail out. Hundreds of the community are peaceable men who arrive with a coat for a blanket, the sagebrush for a pillow, asking naught but the chance to work for a living. But scores are renegades with the future heavily in forfeit, the past beyond propriety to mention, and the present dedicated solely to seizing and holding the chair in the warmest corner. Resident deputy sheriffs are tolerant and their chiefs stay away.

A trip through the taverns by night discloses a large part of the population in the vicinity of the stoves. A newcomer sits with his head rolling, features contorted and feet in a tub. His friends ply him with brandy, for his feet, boots and all, are in hot water and the pain is excruciating. The traveler has just been brought in from the hills; his extremities are frozen. . . . Brace faro, studhorse poker, freeze-out poker, roulette and rouge-et-noir are well patronized, though the three-card monte men, foreseeing a painful winter, are folding their "Spanish games" and departing. The flashier knights of the green cloth also are leaving for warmer climes. Denizens who remain, waiting for the big times promised for spring, find further amusement in the mild new game of starbuck, which resembles twenty-one; or pedro, which can afford much remorse to the man who drops it on the ace. But anyone looking for a fight can find it, and Main Street is quite a shooting gallery.

The honor of opening the gunpowder ball seems to have fallen to Tom Kirby and Gus Norton.

"What have you got?" Kirby called.

"Two pairs," responded Norton.

"An ace full on kings," said Kirby, and reached for the pot.

"Two pairs of jacks," countered Norton, and gently detached the winnings from his opponent's hands.

Whereupon Kirby reached for his gun with one hand, and with the other hand smote Norton in the face.

Norton was unarmed. He accepted the unanswerableness of the situation and retired from the scene. But he was fond of warmth and lamplight and the voices of his fellows, and later returned.

Kirby looked up and saw him seated quietly in a corner. The self-appointed bad man was in his cups by this time. He didn't remember why he disliked Norton, but he recalled that he much disliked him. "I'm a rooster," Kirby announced, "with a long mean spur. Anybody like to cross this line?"

Norton sat unmoving.

"I'm a mighty bad rooster," Kirby crowed again. "If nobody wants to cross this line, I'll cross it myself."

Still Norton did not answer.

Kirby stepped forward, whipped out his gun and fired twice. Both balls struck Norton's legs. As he went to his knees the attacked man drew the pistol he had fetched and returned the fire. He missed.

The "rooster," discovering that the other also had a sharp spur, thereupon fled the resort, darted up Main Street and took refuge under the bed of the first convenient tent. He could hear his late adversary dragging his painful way up the cobblestones. Cautiously Kirby lifted the edge of the canvas and drew a bead.

It chanced that the rooster had taken up ambush in Recorder Smith's tent, and the justice was in bed and interestedly watching the proceedings. Just as Tom Kirby was about to pull the trigger he felt something chilly and round pressing against his own neck. It was only the mouth of a linament bottle, but it felt like a howitzer. At its cold,

HE BEHELD THE '49ERS

Indian George, who also led the way into Surprise Canyon, standing beside the ruins of Panamint's old stage wagon—1935.

*Drawn by E. A. Burbank.*

End of the Run.

"Louis Munsinger set to rearing a brewery."

*—From an old photograph.*

"Was known as Post Office Spring."

firm touch in the dark Kirby just about came apart at the seams.

Norton dragged his painful way to his shack. Kirby on permission of Will Smith cut camp and vanished. The justice of the peace hung the rooster's six-chambered spur on his tentpole and waited for the next item of business.

It came in the shape of an affair between William Mc-Laughlin and William Masterson. The weapons were knives, the distance the width of a cabin table, and the result a job for James Bruce. Bruce felt that the occasion called for style. He made the perforated Masterson presentable, selected a spot for the solemnities well up Sour Dough Canyon, and requested all Panaminters to attend. A prospective hitch in the funeral—the fact that Panamint had no hearse—was solved by Charley King and his two-wheeled butcher's cart. Thereafter the meat cart, being light and easily handled on steep grades, went into official use whenever an irregularity gave Sour Dough a new occupant.

Travelers from Pioche, after reaching Death Valley's floor, were accustomed to rest beside a tiny stream twenty-five miles short of their journey's end. Here Cal Mowry and A. J. Laswell had concluded to make a substantial halt and grow a crop of hay.

Cal Mowry and Andy Laswell were old companions. They had vowed to share everything life provided, from a fortune to a shirt. But after weeks on the edge of Death Valley they gradually discovered that they could no longer endure each other.

The trouble may have started over something quite trifling, such as the squeak in Cal's boot or the way Andy sugared his coffee. An irritant, like a corn, does not have to start large. But it can grow. Finally Laswell went on to Panamint, looked for an answer to his troubles in the bottom of a glass, decided that he had found it and went back to have things out. He was not mean about the matter. Meeting his partner on the trail, he invited Mowry to unlimber his gun on even terms. Cal went down with a ball in his temple and three in his right arm, and was the second victim of violence to be lowered into a grave.

Laswell gave himself up to Deputy Sheriff Rennie, who for lack of a jail told him to go where he pleased, but Laswell by this time felt very bad about the matter and announced that jail or no jail he considered himself under arrest until his case was terminated. "Self-defense," Justice Smith decided—a finding which could do poor Cal little harm and Andy a lot of good.

"Little" Fitzgerald and Tom Hogan got into an argument in Dempsey & Boultinghouse's place. The dispute waxed so warm that both parties set out to fill each other more full of lead than the jumping frog of Calaveras County. Eight shots were exchanged. The final shot scratched Hogan slightly, whereupon the contending parties shook hands.

Jack Williams went "curly wolf" with such vigor that he drew his pistol in Hays & Phillipay's place, shot out one of the lights, and whirling on the other winged himself in the shoulder. His desire to howl was magnificently gratified.

Jim Buckley took the "man for breakfast" theory literally. He assaulted a man named Redding and bit off the upper half of Redding's ear. The torn ear hurt so bad that Redding didn't wait to shoot. He fled, passed Will Smith's justice court, halted long enough to lodge a complaint and then continued fleeing. When Justice Smith called Buckley for arraignment at four o'clock the same day, the complainant couldn't be found—perhaps fearing that his other ear might be served for afternoon tea.

Dick Robertson and Jerry Sullivan couldn't agree about the relative merits of Scotch and Irish distillations, so they went for their guns. Joe Harris, who was cleaning glasses, with the speed of life-practice dropped behind his bar and didn't break a tumbler. Neither contestant hit the other, though each managed to send a ball between the legs of the Chinaman who was washing Joe's windows. This was so funny that the Scotchman and the Irishman had a glass over it.

Ramón Montenegro was of that proud race who called themselves "Californians"—the descendants of Conquistadores and dusky native maidens who for long years occupied broad ranchos and rich, sleepy haciendas before the hurry-up Yankees came. Young Ramón was therefore a don with a code, and one provision of this code was that a man who spoke insultingly to a woman must be knocked down. In consequence, a careless-lipped Philip de Rouche found himself knocked down. The lady who was cause of it all dwelt in the side lane and was known as "The Blue-Tailed Fly."

Philip de Rouche, though not descended like his adversary from Castilian blood—de Rouche was not very certain what kind of blood he was descended from—also had a code. It was that when a man knocks you down he must be knocked over in return.

So a hand reached in through the door of Siebert & MacDonald's resort one night while four men were at play within. The hand gripped a pistol by its barrel, and with great suddenness its butt was made to spoil the features of Ramón.

With these preliminaries the camp was rather sure it was going to need King's cart, and six nights later the guess proved right. De Rouche had stepped out of Dempsey & Boultinghouse's place when a figure behind a post of its wooden porch made itself known and de Rouche walked into a stream of gunfire. The victim staggered across the street, spun and fell. He was carried into the office of the justice and there expired. Montenegro surrendered to Deputy Sheriff Ball, who decided for once that the extremely elastic plea of self-defense was unjustifiable, and took him into the county seat for grand jury action. But no witnesses appearing, the killer was turned loose.

Winter brought other tragedies. One of them descended on the cabin of Lafe Allen and Heber Schuster. The two old enemies thoroughly enjoyed the storm-bound months, never speaking and sometimes almost hesitating to light their opposing fires lest the heat of one give comfort to the other. When, one day, Lafe sickened.

For a time Heber pretended disinterest. Then he yielded to a pull at the heartstrings and advanced on the sick man's pallet. In his hand was a dark bottle.

It was a concoction known as Perry Davis' Pain Killer. Lafe saw it, and yelled. That yell broke the silence of many weeks, and Heber's tongue loosened likewise.

"Take this," he said grimly.

"I won't!" howled Lafe.

"Ye will!" retorted Heber, and inserted the bottle.

There was a tortured gulping and the cabin resounded with enough vocalizing to make up for all its past speechlessness. It became necessary to press a knee to Lafe's chest to make him take his medicine that first time, and it took more than that to force the second dose. For a bit Lafe responded to treatment and almost seemed to be getting well. But long before the single-leafed piñons began putting out their new velvet tips and the yellow butterflies came searching for the yuccas, Lafe heard the angel Gabriel bugling.

All night, all day and all the next night Heber sat beside his pardner, talking to him, pleading with him. But Lafe was bound for a place where the silence would last and this time he was bound alone. When it was all over, Heber broke down. "He was the finest matey a man ever had. A bit long on argument, but given to spells of quiet that evened up. He was right handy, too, about a stove, and his bread *wuz* good, though I hated to tell him so. 'Course he was stubborn. But even that trait showed signs o' softenin'. Toward the end I had Old Lafe swallerin' his medicine like a baby."

Old Lafe was not relegated to stony Sour Dough. In his final moments he had uttered a wish to be hauled to Bakersfield and expressed to some relatives at Oakland. Heber Schuster did not fail in this last office. He saw his late pardner into a box and the box into a wagon, sat with the driver while it jolted eight-score miles through chilly weather, and at the distant railhead dispatched the departed, duly waybilled. But when Heber returned to Panamint, there were no more sentimental remarks. Old Lafe, it seems, had taken advantage of the opportunity to deal his friend and foeman one last, looping, underhand blow.

"He took," groused Heber, "four dollars' worth o' ice!"

PANAMINT continues tough. Dempsey & Boultinghouse have, for neighbors, a hurdy house that clings to the ascending cliffside at the rear. Its dance floor overlaps the ceiling of their public room. To the furniture of those realms above, a cord is attached and led down through a hole; to the lower end of this cord is affixed a "bug" made of corks, shingles, nails, toothpicks and chicken feathers. When the bug begins to cavort, spectators below are given a fine fright and other customers come running—which provokes, Dempsey & Boultinghouse find, much business.

Jim Honan of the Silver Glance mixes up a concoction of alcohol, jamaica ginger and Perry's Moth & Freckle Lotion to shine the mirror. When a newcomer makes disparaging remarks about the establishment's wet goods, comparing them with those he has nearly been done-in by in other camps, the glass-polishing mixture is recalled and shoved his way.

"What's that?" asks the hardy customer.

"Panamint cocktail. Very searin' to the vitals."

"Put something in it to give it wires—I'm from Pioche."

Jim Honan, who is an expert, adds bitters and a dash of pepper sauce, and possibly a bit of cut plug. "How d'ye feel?" he asks a moment later, a trifle anxiously.

The pallid customer grips the counter. "Fine!" he gasps. "That there was likker. Put in some teeth from a circular saw and durned if I wouldn't *feel* it. . . ." Two residents are doubtful of the "scorpion juice" discovered in an owner-less bottle, so they call to a third, who is admiring their find, and offer him a swallow. The human guinea-pig grows ashen, sags in the knees and shakes himself with a bellow.

"How was it?" demand his hosts.

"Just right, gen'l'men, just right! If it had been worse it would a' killed me, and if it had been better I wouldn't a' got none." . . . One inventive mind devises the trick of sitting down alongside the barrels. When no one of authority is looking he turns the faucet and lets the fluid run into his boot-top. In good time he shuts off the flow, walks out and is soon behind some shelter dividing with a friend. A joke on the bartender is always the richest kind of a joke, any time, anywhere. . . .

A mystery shot, origin unexplained, winks out the light over Nick Perasich's head in the French Café. Twenty men are available next day who swear they felt the bullet part their hair. What annoys Perasich is that the missile imbedded itself in a head of cabbage, and cabbage leaves up there in December are almost as valuable as greenbacks.

By this time Panamint is getting a reputation for lawlessness, and a report reaches the lowlands of six men dying in one Homeric fracas. The tale is a fabrication inspired by the ejection of some citizen from a place of strong waters so forcibly that he just can't rightly recall all he saw en route. But the legend reaches Los Angeles and is printed in its newspapers.

Gambling goes on for high stakes. Most of the snowed-in miners are poor men, but a pot can take on size after a little concentrating. Somebody reports a hand that looked so pretty that Ned Reddy, its owner, backed it with $1000. His opponent stayed and drew. What he drew led him to bet $4000 more. When the pot reached $10,000 they had a look at each other, and sixes and aces swept the cloth clean.

A dicing game currently popular in San Francisco and Virginia City is also imported and kept running day and night. The game is played with four cubes. Anybody can join in by tossing four bits in the pot. The pot goes to whomso rolls the highest number. The house, for its cut, exacts two bits every time four-of-a-kind are rolled. In return, the house settles arguments and provides refreshments; in spite of which trouble and expense, the profits are good. . . . Down at Dempsey & Boultinghouse's place, Jim Bruce quietly checkmates a trick familiar enough to a Piocher. He is asked to take a drink, the glass being offered on a tray, and the tray being held high and pressed against his breast. With his view thus interrupted, a hand surreptitiously reaches forth to alter the cards in the faro box. Bruce accepts his glass with one hand, covers the box with his other, and the game goes on.

Up Sour Dough way a young man named James Earle, who has exhausted his credit, and down-canyon a young lady named Featherlegs who has extended credit too widely, alike find continued cold and hunger unbearable and settle the matter with as little bother to the community as possible. Jim Bruce attends to the obsequies and quite a crowd turns out, chiefly to shed regretful tear for the little gal so generous. . . . Yet, with dry snow blowing in great clouds off the mountains, it is pleasant to sit indoors with a hot Scotch and recall how the boys died of heat last summer down on the Panamint Valley and Death Valley floors. Anybody who was here before July is a pioneer, and commands respectful attention.

Wrestling, "collar and elbow" style, is popular, some of the cyprians being not averse to taking on an opponent, male or female, for a bout on the floor in consideration of a purse and side-bet. Recitations by the gifted are in maudlin favor, in especial request being "Somebody's Mother" and "Look Not upon the Wine." There are faces that grow seamed with doubt when a high voice sings, "Is My Darling True to Me?" An occasional cyprian, full of fighting whisky, starts to clean out the dove-cotes and provides grand entertainment.

So it befalls that in the lower end of town, after an evening of numerous callers, silence embraces the house of Martha Camp. Into the outer darkness departs Mr. W. N. McAllister, proprietor of the Snug Saloon; Mr. Edward

Barstow, night watchman for the Panamint *News;* and Mr. Barstow's particular friend Jim Bruce.

"Come in," says Mr. McAllister to his friends at the doorway of his own place of business, dark and tranquil under a midnight moon, "and have a little something."

A little something runs into a little more, during which process Mr. Barstow loses count. He also, it gradually occurs to him, loses Mr. Bruce.

Back at the house in Maiden Lane, Martha Camp is preparing for what is left of the night. She unpins and places on her dresser the long earrings, bracelets and brooch with which she adorns her charms. Beneath the edge of her bed she places her high black shoes. She removes and hangs up the flounced and beaded gown. She removes her slip and her several muslin petticoats. She removes her long cotton stockings, her pantalettes and her corset cover.

Then struggle ensues. Around the cupping top of the garment now revealed there runs, like battlements to a fortress, a scalloping of lace. This, and the lover's knot of ribbon that terminates it, are deceptive frills. The inner substance of this structure is iron and whalebone. Their remorseless purpose is to constrain and mold. If the upper half of this container provides area for an abundant matron, at its waist there is scarcely wriggling room for a nymph. Daylong, above the bell-like opening, there has been freedom for the neck and arms, and somewhere down below there has been freedom for the knees. Between bust and thighs all has been caged restraint. This castle was brought closed at dressing-time by Martha's muscular handmaids and so it has remained, gated at the front by six strong snaps.

But now after a tussle she steps free, a Venus in full-length underwear, and sets the fierce machine aside. Through the night it shall stand there, erect by its own might. Still attached are the two appendages which it supported at the stern all day. One, a contrivance of reeds and wire, and the other that rode above it, a horsehair-stuffed bag, betray how this wonderful race build out their fascinations at the back.

Making sundry last divestments and shaking down her hair, which she twists about an array of kid-covered curlers, Martha permits the attending Sophie Glennon finally to blow out the lamp.

A masculine pair of French-leather boots already stand beneath the bed. The owner of this pair makes room at her vigorous shove. Save for the snores of Mr. Bruce, who owns the French boots, all becomes quiet in the house of Martha Camp.

All is quiet for an interval. Then there comes a pounding at the door. Mr. McAllister, who is retiring in his own cabin, and Dr. Bicknell in his quarters on Main Street opposite the entrance to the Lane, will testify later that they both heard a man's voice shouting. Questioned as to just what they heard, witnesses will state that the voice was proclaiming somebody a son of a shooting word, and announcing to the sleeping town, the Panamints, and the setting moon his intention immediately to bust in, gun smoking, and clean out the joint. They also will say that they heard Martha Camp scream, "Go away!"

Dr. Bicknell perceives he will soon be needed, and rises to don his trousers.

The uproar outside, and the pleas of his lady beside him, rouse the easily waked Mr. Bruce. He conceives that the sonofabitch being so loudly referred to must be himself.

Mr. Bruce is not a man who goes outside for trouble. He

much prefers to let it come to him. He summons Sophie Glennon, instructs that vestal to light her lamp again, and directs her to stand with it by the bureau where the beams will fall just so.

Barstow bursts in, as he has been promising. Martha Camp shrieks a warning—he is, after all, a steady customer. The bright yellow cone of Miss Glennon's lamp falls warmly —welcomingly. Bruce remains in shadow. He permits Barstow, gun in hand, to advance two steps. . . .

Directing Sophie to lift the lamp again so that he may see under the bed, Bruce reclaims his shoes and hastens to fetch Dr. Bicknell. He meets that gentleman hurrying up the hill, professional kit in hand.

Barstow is carried into Bruce's own cabin and laid on its bed. Bruce then goes for Dr. Wells to come and assist Dr. Bicknell. Though they remove a leg, the two medical men are not able to save the rest of their patient, who dies exonerating and forgiving his friend and rival.

Barstow's funeral is the most elaborate of any that have so far taken place in Sour Dough Canyon. Bruce is meticulous as a director. Martha Camp is visibly affected. Sophie Glennon trembles like a leaf throughout the ceremony, and though lamps burn late o' nights in Martha's place thereafter, nothing can induce Sophie to handle them.

BUT PANAMINT was not without graces. A white girl baby was born in one of the rude cabins in mid-winter and the populace sent her a case of champagne and united wishes for a life of "bay trees and clover." True, in a cabin down-canyon a Mexican miss had got the start of her by three months, but to that event there had attached no such social significance. . . . Buffoonery beginning to pall and shooting scrapes being after all an amusement with limitations, somebody suggested getting up a spelling bee. This innocent hilarity was currently sweeping cities, towns and villages everywhere. The camp's most confident spellers began to talk of a tournament, the winners to challenge Independence for the county championship, and the thought occurred that it would be fine to let the seat of county government be the prize. Independence declining the challenge for something that it could not benefit itself by winning, but could only risk losing, a community-wide spelling bee was organized in Panamint anyhow.

Such an affair is conducted as follows: the entrants buy places in the line at two bits a chance, withstand bombardment from the lexicon as long as they can, and when overthrown go to the foot of the line and work up again with a payment for each defeat; the proceeds going to gentle charity.

Miss Donoghue's Wyoming Restaurant, for its eminent respectability, was chosen as the arena and cleared of its long tables, and Editor Harris as the camp's exponent of the belles lettres was installed as moderator. The beneficiary was to be an ailing miner who had over-tarried near a short fuse.

All went well and quite a pot was collected before Moderator Harris showed that unfortunate lapse of judgment. He had hurled "heliotrope" and "hellbender" at the contestants without serious results; but when he leveled a finger at Jerry Sullivan and challenged with "hoarhound," it took several men and a table leg to restore order and Jerry stamped out vowing it a lie and a libel, and anyway a word that any man of the world could spell with his eyes shut. This matter became the camp's whoopingest jest. Later, when Rob Govan, a youthful carrier for the *News*, entered a resort with his papers, Jerry pursued and felled him with a blow from his revolver-butt, thereby restoring the prestige of Colt over Webster.

"We desire to have it understood that our employees," stated Editor Harris, "have no voice or responsibility in our editorial opinions."

But the fleshpots of civilization were arriving. Western Union promised a telegraph line by spring. Rufe Arick petitioned the Inyo supervisors for a franchise to build a cable-tram from bottom to top of Surprise Canyon and there was likely talk of a railroad soon raising its steam voice at the lower end of the gorge.

A massive billiard table, purchased at San Francisco and shipped by sea to Wilmington, was reported on its way up the passes and across the wastes behind a sixteen jerk-team guided by an infallible skinner. A freight wagon arrived from Columbus, Nevada, heaped high with bedding for a lodging house. Louis Munsinger put down a well, struck good water and set to rearing the stout log walls and chimney of a brewery.

And into a brand new structure moved Neagle's Oriental. His $10,000 worth of fixings included space for the oncoming billiard table; a black walnut bar; wainscotings of Inyo pine finished to look, at a distance of ten feet, just like real oak; walls above the wainscotings done in selected poses of the female form by a Los Angeles sign-painter in the best American school of popular art; and two cardrooms, separated from Joe Harris' Occidental next door by a bullet-proof wall.

An occasion of almost equal importance was the opening of Fred Yager's establishment. Scorning local stone, the

Dexter was built of milled lumber whose hauling took longer than would now be required for a trip from China. The Dexter's square front was supported by "numerous and elegant brackets," as the *News* described them, and it welcomed the town through two pairs of transomed doors. While shy of paint on the outside, the interior was provided with satin-gilt paper and two chandeliers of four lamps each, hanging from a fourteen-foot ceiling. Behind the bar, widely spaced, were two side-chandeliers, and the wall between them was intended for Panamint's proudest jewel serene. Senator Stewart in his mansion at Washington might have the biggest pier-glass in the United States. But Fred Yager was going to have the biggest mirror in the Seven Deserts.

This noble piece of plate, seven feet by twelve, after a voyage around the Horn started its final journey from shipside in February. Its progress was reported day by day by arriving travelers who had passed it on the way.

Behind many mules it made its way into San Bernardino Valley. It surmounted Cajón Pass and negotiated the double fording of the winter-full Mohave River. It lurched out across the Mohave plateau and tossed back the reflections of mountains, mirages, alkali flats and stormracked skies. Advance riders reported its safe arrival at the bottom of Surprise Canyon. It was next reported proceeding through the Narrows, where the cliffs of that gulch came close, and then it was passing Jacobs' mill. The remaining mile was a triumphal procession—the crowning incident to Washington's Birthday celebration.

So it was that the plate at last reached Mr. Yager's wide-opened double doors. Admiring hands plucked it down from the wagon, while the wayworn teamster accepted praise with the air of one who had labored well. Respectful feet marched the big mirror toward the threshold.

But some of the feet were tangled by pre-celebration. One or more pairs of them stumbled. There was a yell from Mr. Yager and his barman Bob Peterson, whoops and bedlam from the onlookers. A shivering crash was followed by sudden silence.

The space between the waiting side-chandeliers was thereafter filled, as the *News* with forced cheerfulness described, "by five elegant chromos, representing the delightful and romantic scenery of the Rhine, and so disposed as to present, to our taste at least, a much more pleasing and satisfactory scene than that of a fellow with a dirty shirt standing before a magnifying mirror."

At the extreme upper end of town Miss Delia Donoghue heard sounds in lower Main Street, produced by burning pistol-powder, which showed that Washington's Birthday in spite of this disaster was still being observed.

Miss Donoghue never ventured far down that road, being a respecter of deadlines and anyway usually occupied with her range, and today she was busier than ever. For tables, stove and counter had been removed, a platform set up, many sandwiches prepared, and now the musicians—harp, flutina, concertina (played by Editor Harris) and Professor Martin's fiddle—were tuning up. If there were only more ladies in camp! Miss Donoghue for the dozenth time counted over all the genteel prunellas which were available for partnering the miners' heavy stogies on her tallow-waxed floor. Just sixteen pairs. Down-canyon, of course, there were other "ladies." But Miss Donoghue's face went bleak at the thought. "Thank Providence," pursed Miss Donoghue, "the road up this way is too steep for *those*."

So now while heavy soles and light threaded the mazy, and Professor Martin called the square-dance figures, the polished lamps shone warmly and the piles of sandwiches vanished. Long after the last rustling dress and flannel shirt had bowed adieu, Miss Donoghue stood happily amid the debris of the night and reveled in the memory of it.

Panamint might be rough. Panamint might be tough. But Panamint was making progress. Truly, as Jack Lloyd the stagedriver had been moved to remark when he looked in upon the scene, "There's many a noble heart beats beneath a tattered pair o' pants." Panamint had shaved its chins, cleaned its boots, produced toilettes "elegant and tasteful," and with style and no shooting had flung its first ball.

Though, looking up at the star-studded sky, Miss Donoghue realized with a start that the comet had long since departed.

# 17

# THE ENTERPRISING FIRM OF
# SMALL & McDONALD

HEWING tunnels and grading trails did not appeal to Small
and McDonald, two of the important claim-owners. They
came down occasionally from their cabin high in Wild Rose
Canyon and looked the young town over. What particularly
interested John Small and John McDonald, and had interested them from the beginning, was that small stone building whose ground plan was taking shape halfway up Neagle's
Block.

To start a bank in a brand new, wide-open mining town
you need a few dollars in capital, or the reputation of having
it. You need faith in your fellow men. You need, for balance, a cashier who hasn't the slightest taint of such faith.
And you can do with a banking structure stronger than a tent.

Louis Felsenthal had arrived from San Francisco late
in November by Buckley's stage line, and in one month a
rising stone building had "Bank of Panamint" over it. Felsenthal was not exactly an innocent. He owned merchandis-
ing establishments at Hiko and Pioche, and it was in front
of the doorway of the latter place that Dave Neagle had
loitered with a gun, looking up the sidewalk for Jim Leavy,
when Leavy emerged instead from an alley that ran alongside the store and handily surprised Neagle and a companion.

"The shooting," said an eye-witness, "commenced at once." It was in consequence of Mr. Leavy having proved impossible to kill, and the experience of occupying the same cell with him during his convalescence, that Mr. Neagle had since moved on to his present sphere of action.

By this and other incidents close to his doorstep, Louis Felsenthal knew the ways of silver camps. But he seems to have been an incurable optimist. He disappeared down Surprise Canyon and presently came back with a skeptical-looking cashier and some hefty canvas bags.

The cashier, Morris Eissler, cast a fleeting glance over the town and found little to change his preconceived notion of it. Chancing to observe John Small and John McDonald standing before Neagle's place, his look became still more glum. John Small was a short man as suited his name, though otherwise built along the lines of a butcher's block or a railroad round-house. His companion was thin, rawboned, and wore his pants pockets high off the ground. In matters requiring conversation it was Small who did most of the talking, though McDonald from a background position usually voiced the decision and was the dominant member of the firm. The pair went in for heavy dark facial growth, and the eyes above this hairy cover were alert and hard.

Cashier Eissler had a professional memory for faces, and he recalled those two; though whether it was at Eureka, or Austin, or Pioche . . . Meanwhile there was work to do. He almost groaned, and permitted his knees to sag slightly, as he received the canvas bags from aloft.

Curiously enough, Charley Haines the driver afterward mentioned that he had found little difficulty in tossing the sacked mintage down by one hand.

The experienced appraisers Small and McDonald had

made it a point to be present at this opening. John McDonald with difficulty had restrained his ally from halting Haines' stage in the canyon and relieving the bank of its funds then and there.

"You're not thinking straight at all, John," reminded the taller partner. "A new bank is like a jackpot. You must let it grow. You take it over now, and what've you got? Just Felsenthal's ante, and I'll bet that's in iron washers. But you leave it be awhile, like any other pot, and what else is in it? Everybody's chips. Now, John, think farther. How does Felsenthal or any banker figger to make his bank pay? Why, by takin' up the public's own money and loanin' it back. That'll be where we come in, understand? We'll step up and borrow what we require. We'll borrow real frequent and regular. But we've got to take care of this bank if it's to take care of us. So, John, step easy."

Thereafter the bank's relations with the sweet-scented buds of Wild Rose Canyon were intimate if not two-sidedly enthusiastic.

"How's she goin'?" John Small would inquire from time to time with proprietary interest. John McDonald, saying nothing, would be present at his shoulder.

Cashier Eissler would move his stacks of gold and silver —customers' contributions—a little farther from the window. "Slowly—very slowly, gentlemen. Times are hard, what with world-wide demonetization on one hand, and the local difficulties of reduction and amalgamation on the other. Very hard indeed." The cashier would sigh, his expression calculated to wring the heart.

"Keepin' the capital tight in that there vault?" The question would be accompanied by a heavy wink.

"Safe as in God's pocket. We never take *that* out."

"No-o, I reckon not. This mountain air might rust it." The sally would be followed by hearty laughter from the visitors, a poor effort at a smile by the cashier. Then: "Well, guess we'll have a loan."

"Certainly, Mr. Small. A couple of fives? Demonetization, you know, and the difficulties of reduct——"

"Raise you, Mr. Eissler. Some of them twenties will be just fine. Well, take good care of the bank."

Cashier Eissler would look with rue at the revolver he kept beneath his counter and with venom at the departing backs. It *was* in Austin that he had seen this duo; he remembered now. And the recollection convinced him that a little off the counter from time to time would be infinitely better than a lot from the vault. Not that the contents of the vault were so valuable to the absentee Felsenthal. But a public inspection of their make-up might be unpleasant for his present-in-the-flesh cashier. So Cashier Eissler laid out his loans as his better prudence dictated. And up and down Main Street the borrowers would go, treating the boys, and leaving on each bar all the change from a gold piece just in case anybody would like another. As everybody would, this gesture made them popular along that rough mile.

Occasionally the mood for kindly doing led them to finer flights. A story was in circulation of their going straight from the bank to the cabin of a widow who had lost her husband in a mine disaster, and assuaging her grief with a stack of double eagles.

By this time Small and McDonald had come to be regarded as a pair of uncouth but golden-hearted Robin Hoods who robbed the rich to treat the poor and were the town's chief ornament. As they wore big guns with the grips slung forward, argued fiercely—or so John Small did—the comparative merits of thigh-holster, arm-holster, and double-cross or two-handed grab from the belt; practiced gun-fanning, which is performed in a kneeling posture with the trigger pulled back and the hammer of the revolver struck with great rapidity by the stiffened palm—very effective in a tight spot; shot tin cans and kept them rolling, and strode Main Street with an independent air, there developed quite a legend about them.

The facts underlying it seem to have been these: the pair were up in the hills cutting nut pines for their winter firewood when they were interrupted by a Panamint Indian who —not very originally—wanted something to eat.

Panamint Indians on a search for something to eat had been dwelling in that hard range since the prehistoric start of things, but this native had the chief trait of his race extraordinarily developed. His reputation was not of the best. He lived beyond the divide and was suspected of having done away with several white men. John Small, somewhat short-tempered by his labor, told him to get busy with an ax and they would feed him at dinner time. Indian Jake was not disposed to work, still less to wait. He wanted dinner right away.

John Small did not think this was showing the right respect, so knocked him down.

The Panamint first citizen got up yelling threats in several dialects and disappeared up the ravine. When Small and McDonald came down to the town again they found the air filled with rumors of an impending Indian uprising said to be taking shape on the Death Valley side.

The two Johns promptly announced that this was their private fight. Advising the rest of the town to keep cool, they strode off up the valley with quite an armament. They climbed the stone curtain between Telescope and Sentinel Peaks and dropped out of sight. So much was definitely known. Two days passed, replete with rumors; then the Indian fighters returned.

First report, delivered over a preliminary hooker of brandy in the Oriental, mentioned coming upon and putting to rout two Indians, one of whom was Indian Jake and the other his squaw. Succeeding reports, delivered over successive hookers, jiggers and schooners, developed the tale a trifle. There had been six Indians, said Small—eight, corrected McDonald—all heavily armed and done up in war paint, who had tried to ambush the pair on the trail and who had been handily extinguished for their pains.

Toward midnight, under much pressing, the stalwart pair at last yielded what they assured was the full and truthful narrative of the fracas.

On the edge of Death Valley they had come on a stone corral containing something like fifty Indians, all worked up into a rage and led with fiendish energy by Indian Jake, now tuned up into a combination of Tecumseh, Cochise and Captain Jack. The white men had wormed their way up to the enclosure by nothing less than the adroitest cunning and had engaged its occupants in a gun battle of many hours. When the smoke cleared away, almost the entire enemy force was stretched on the ground and the rest were disappearing like scared lizards for other crannies. Toward morning, if the attention of the audience still held out, the list of casualties must have grown to proportions truly bloodcurdling; but it is more likely that the listeners were sunk in stupor and the heroes themselves a bit weary after so much carnage. At all events, the Battle of the Stone Corral added many inches to the local stature of Messrs. Small & McDonald.

After resting and recovering from this saga of mortal combat, Small & McDonald began casting about for something more to do. They were incited to this by a somewhat uncomfortable feeling that the Battle of the Corral had its belittlers, not to say its out and out doubters, particularly as Indian Jake was once more around and not even a square meal could induce him to remember the tremendous set-to or the army he had led.

True, the analysts kept their opinions to themselves, or unleashed them only in places not patronized by Small & McDonald, but there the situation was. Narrow gorges can toss rumors as well as echoes. Small & McDonald perceived they needed something that would reaffirm their prowess.

What their searching eyes lit upon was just the ticket. It offered them a chance to go against overwhelming numbers, in practically direct view of their fellow-townsmen, and —though Small & McDonald probably resisted the thought —at not slightest risk to themselves.

Down the grade, close to where the real pitch for the desert began, was a quarter on elevated ground occupied by two or three score noisome hovels. These were the dwellings of a few hundred Chinese, who were wintering in close and happy confinement after a hard autumn's road-work. The Confucians were running their fantan and narcotic dens without external assistance and were molesting nobody but each other, but their presence was a threat to white men's employment and a standing racial annoyance to the Swedes, Irish, Dutch, and all the other one hundred per cent Americans of Panamint except, possibly, the Panamint Indians. There had even been mass meetings about the matter.

Small & McDonald took over responsibility for settling the issue.

New Year's Eve was well past the hour of high carnival and the Swedish, Irish, Dutch and other one hundred per cent drunks were falling asprawl in their favorite corners when the brave pair belted on their guns, collected an armful of rocks and as much of an admiring audience as could stagger, and advanced on the slumbering hovels, dark now save for the fitful gleams of opium lamps.

Under the influence of the poppied pipe-bowl the form relaxes, the features turn waxy and strange visions fill the inward eye. At first the bombardment of rocks on their doors and roofs broke through to the senses of the celestials as something remote and even rapturous. But the missiles were heavy, Small & McDonald were resolute of arm, and some of the porphyry hail plummeted straight through.

That broke up the dream. With arms clutching apparel and queues streaming, the Asiatics piled out the rear doors of their cabins as Small & McDonald, with stones, oaths and shots, thumped and stormed at the front.

Down the canyon fled the woebegone coolies, stumbling in chill starlight that was just beginning to give way to dawn; bearing their caught-up belongings in pole-slung baskets and dragging or driving their female chattels over the rough roadway. Down the canyon as far as Pott's Station the shouting, shooting pair of Caucasians chased them. Then, the fugitives well provided with momentum, they went back and put torch to the abandoned cabins, and returned to Cervantes & Perasich's French Restaurant for a good-humored New Year's breakfast.

Out upon desert paths the expelled Chinese went fleeing, and for many a thorny mile the route lay ahead. Snow, rain, and generally bitter weather helped to make it a via dolorosa whose end, no doubt, some of the exiles never won.

Up on the shelf of Surprise Valley the camp awoke to hear the robust tale of the small hours' doings. Its swash-buckling heroes were considerably feted. All in all, the camp didn't have such a jolly hour again until Archie MacDonald's saloon burned up.

# 18

# THE COMING OF THE SHERIFFS

EUREKA, off in middle Nevada, was in this period a town of two important hotels, two newspapers, many mines, much hustle, and about five thousand people. It was the base of supplies for Hamilton, forty miles away, and all the White Pine country; for Ward, 100 miles; Pioche, 190 miles; Tybo and Belmont, each 100 miles distant; and in particular it had much in common with Austin eighty miles west. To all of these points radiated its bustling stage lines.

Life was so full, pulsating and frequently brief in this seat of Eureka County that its chief organ the *Sentinel* was able to remark: "It is a fact not generally known that Eureka is provided with seven separate and distinct graveyards. It is but a little over six years since the first graveyard was started in Eureka, and now there are attached to it seven comfortable and well-appointed hotels for corpses. What an evidence does this fact present of the marvelous progress made by our town in the short space of six years! The future of a town with seven graveyards is assured."

To get into this enterprising community from other points of civilization, one took the overland railroad to Palisade, a station set deep between towering cliffs on the Humboldt River, and there transferred to a stagecoach that wound southward through sagebrush for ninety miles up steeply rising

Pine Valley. The narrow-gauge branch railroad was under construction but not yet completed. Returning from Eureka, one again climbed aboard a swaying Concord, and if experienced in desert travel one selected the stage that traveled at night.

Such was the policy pursued by four passengers, one of them a lady of multiple petticoats, on December 3rd, 1874. The stage that left the Jackson House contained, besides these travelers, an express box rich in treasure lately torn from Ruby Hill. To see that this bullion from the great Richmond mine reached the railroad junction intact, Wells Fargo's armed messenger Jim Miller climbed aboard. Movement over the road had been rather peaceful of late, so Jim rested his shotgun against the back of the seat.

"I'll stand Maria there," he told the driver, "until we pass Diamond Well."

Diamond Well was an unimportant sort of a place if you were traveling north, in cool starlight, behind fresh horses; though its bucket, windlass and long rope could have real significance if the hour was midday and you were coming the other way.

The stage jingled along. Maria, the sawed-off twelve-gauge, stood modestly behind Jim's elbow. There was a bit of nip in the air. No need to stop for water. That was just this stage-load's good luck, of a sort; for the bucket was floating on the surface of the water forty feet below and, unsuspected by the travelers, there was no longer any way to reach it.

About a hundred yards short of the well a black object took form in the obscurity on the right-hand side of the road and a voice said curtly: "Pull up!"

The driver was of the hardy breed of Nevada stagemen. He didn't like orders from anybody, however politely phrased, and he cared still less for sudden commands from strangers. He seemed to have difficulty bringing his several horses to a standstill, at the same time nudging Jim Miller hard in the ribs.

"None o' that!" ordered the waylayer sharply as the messenger's hand moved toward Maria. From the opposite side of the road came a similar warning.

So there were two of them.

The driver was not through hoping for a settlement by gunpowder. He reached for the express box as directed, but pretended it was wedged in the fore-boot. While fumbling, he found opportunity again to nudge Messenger Miller. Jim Miller knew what he meant. As the driver up-heaved with the box, creating an instant of distraction, he made grab for his gun.

There were two belching roars. The messenger's salvo soared out into the general sagebrush. A robber's answering gun hurled its clot of disablement straighter.

"Now, Jim! T'other barrel!" yelled the belligerent whip, kicking loose his brake. In spirit he was blood-brother to that Driver Peters who had behaved exactly this way on the San Bernardino road four months before. As in that case, the horses now swept forward at a frenzied gallop and their driver ducked. The pursuing hail of shot from the robbers' weapons pattered on the leathern hind boot of the stage. The inside passengers were wonderfully flat on its floor.

But Jim Miller did not respond with t'other barrel. Maria had spoken her piece for the night. Her master was slumped in his seat. The driver just had time to clutch him with one hand when the horses went down in confusion. They had been brought to earth by the stolen well-rope, which had been stretched taut across the road.

Thoroughly covered this time, the driver delivered up his express box. His lead-horses were gone for the far places and his swinging-pole smashed. An hour later the defeated coach crawled back to the Jackson House with its wounded messenger bandaged in the woman passenger's torn-up underskirts.

Miller had recognized his assailants. They were the pair who had hung around the neighbor town of Austin both before and after several painful occurrences on the roads out of that town. A sheriff's invitation yelled into the doorways of Eureka's all-night taverns rounded up a pursuit party. Dawn showed a clear pattern of the fleeing horsemen's tracks. But the rising sun brought only difficulty. The tracks moved out into the sagebrush in wide spirals, and the spirals crossed and recrossed each other in baffling fashion. The only certainty to be made of the matter was that the horsemen were moving north. Finally the trail vanished altogether on rocky ground.

But John Small and John McDonald were not headed north. After cutting each other's tracks to mix up pursuers, they had dismounted, tied sacking about their horses' feet, and then cast a long loop southward. A week later they were cozy again at Panamint.

This time Wells Fargo was looking for the holdup men with real energy, scattering reward posters over a thousand miles of countryside. "Send a copy of the indictment or complaint against Small and McDonald for robbing the Eureka stage. They have been captured," wired the governor of Oregon. That seemed to settle it. But photographs of the arrested suspects, mailed to Eureka, proved that the men held at Eugene City were the wrong pair.

Rewards were increased. The blackest type obtainable

in mid-Nevada was used to tell in what value the express concern held Small & McDonald, messenger-shooters and nine times depredators.

One of these posters was mailed to Panamint. Jim Brown hung it prominently by the doorway in Harris & Rhine's general store and stage office.

An individual paused before it one morning and read every word with rising indignation.

"Two thousand dollars apiece for Small and McDonald!" snorted John McDonald when he reached the bottom. "Why, that's blood money! This here will sure make John Small mad if he comes along and sees it"—and he ripped the poster down before several pairs of startled eyes.

In January an unnamed citizen wrote from Panamint to a friend at Eureka, who turned the letter over to the *Sentinel:*

"The gang of cusses who shot Jim Miller are here, or a little way out of this place, at Rose Canyon. There is a whole nest of them. I have seen them in town several times, and I think it would be well to let Jim know of it and have the W. F. Co.'s detectives come around, for they make their boasts that one hundred men could not take them. Parties who have recently arrived from that place confirm the above, and with a reward of $4,000 for the capture of the men it seems strange that someone can not be found with 'sand' enough to tackle the men and get the coin."

At that, Archie MacDonald, proprietor of a Panamint public house often patronized by his namesake, took his pen in hand. An alibi for the Eureka affair might be difficult, but the reward poster also cited the Austin–Battle Mountain hold-up of August 13th and here, perhaps, something could be done for such estimable customers. He addressed Sheriff Emery of Lander County:

"My main object in writing to you is to inquire of you if there is not some mistake in charging Small and McDonald. There are locations made by them here in the month of August, and the recorder says they made them in person. I have conversed with Small on the subject, and he says that if he had money for the expense that he would necessarily be put to, he would return to your part of the country and have an investigation. He has very valuable mining ground here and says that as soon as he can dispose of it he proposes to do so. These men are in town frequently, stopping for weeks at a time. Everyone almost knows them and of the reward, but believe that the men could not have been here and at the scene of the stage robbery at the same time."

"They were identified at Austin and again at the shooting of Jim Miller," retorted the *Sentinel*.

Three months of this, and Sheriff Gilmore of Eureka County, Nevada, concluded that he could no longer refuse the challenge to the journey. His conclusion was helped on by the arrival of a communication from Panamint of unrevealed contents, and another from San Francisco. He tried to take Sheriff Emery of Austin with him, but that official of Lander County had lost interest in men who were trying to make a new start in a stronghold a mile high and several hundred miles away. He gave Sheriff Gilmore the authority, however, to act for both. The good Gilmore thereupon departed for his men alone.

"We fear," said the Austin *Reveille*, "he will have some difficulty in effecting their capture, as the population of that section is largely made up of men who are naturally in sympathy with criminals, and who will place every obstacle in the way of the officers of the law."

At Independence, Gilmore picked up J. J. Moore, sheriff

of Inyo County, and after a bit of argument the two rode for Panamint. The sheriff of Inyo was not wholly pleased to find himself going; there was usually trouble enough at home without riding so far afield for it. "These here bad men," he explained, "are like chuckawallas. When you chase 'em they dodge and run until they get up into the rocks, and there they swell themselves up so that even if you lay your hands on 'em you can't pull 'em out. Yonder in the Panamints these fellers Small and McDonald have swelled themselves up something tremendous." But the sheriff of the local county was prevailed upon to go when a tall, blue-eyed man in a dark blue suit descended from the stage just arriving from the south, and further helped Gilmore to explain his mission.

Jim Hume was Wells Fargo's chief special officer. Shotgun guard and sheriff's deputy, he had been riding stages over the battle-torn mountains for a decade, was noted for his swiftness of hand, and had held the post of chief detective of the express house for about a year. Over the next two decades Jim Hume would prove nemesis to more than five hundred stagecoach thugs; he was to become one of the great pacifiers of the West. There was power in those cool blue eyes. Though with Jim Hume, as with the house he represented, the motto was Business First. Catching robbers dead or alive was a paying sport, but getting treasure through or getting it back when lifted was the prime consideration.

Arrived at Surprise Valley, the trio went into conference with a big-bearded figure who seemed much at home in that environment, and whose eloquence had often been retained by Wells, Fargo & Co. To the meeting also came two figures who rode in from Wild Rose Canyon. Their heavy guns were worn low on the thigh with the grips well forward. These they lodged with Dave Neagle before proceeding to a meeting in the little back room.

"Gents, these here are John McDonald and John Small. They rate among our leading citizens. Mr. Small and Mr. McDonald, shake hands with Sheriff Moore of this county," said Neagle.

"Pleased to meetcha," said the leading citizens politely.

"And," said the master of ceremonies, "this here's Jim Hume."

"We've seen him before," said John Small. "We generally aimed to let him go right by."

"This is Sheriff Gilmore," went on Dave Neagle, "over from Eureka. He's ridden quite a ways to talk with you boys."

"Sorry we didn't know he was requiring us," grinned Small.

"And this last is Senator Stewart, whom you know. Well, gents, I guess this completes the party."

"John McDonald and John Small," said Gilmore with formality, "I charge you with holding up the Eureka–Palisade stage on the night of December 3rd and stealing Wells Fargo's treasure box, besides shooting Messenger Jim Miller."

"That is all right," responded John Small easily.

"I also understand," said the sheriff, "that you are prepared to sell a valuable claim known as the Ophir to the Surprise Valley Company."

"That's right, too."

"John McDonald and John Small, will you come back to Eureka willingly, or shall I proceed in the performance of my duty?"

"How much does the warrant say we owe Wells Fargo?" asked John McDonald briskly.

"Four thousand four hundred and sixty-two dollars," spoke up Bill Stewart as attorney for Wells, Fargo & Co.

"Four thousand four hundred sixty-two dollars and sixty-four cents," corrected Hume.

"What are we offered for our mine?"

"Twelve thousand dollars," volunteered Bill Stewart in behalf of the Surprise Valley Mill and Water Company.

"A third of that goes to Dave Neagle for our grubstake. Wells Fargo gets $4,462.64. The rest belongs to John and me. How does that leave the account?"

"Squared completely," said the big-bearded attorney for the express company. "Wells Fargo's interest in its treasure boxes is wholly practical. Suits the Surprise Valley Company, too," he added as co-proprietor of that enterprise. "Mining claims don't care who locates 'em or who sells 'em."

"Suits me," said the sheriff of Inyo with enthusiasm. "I never lost you fellows anyway, so I never tried to find you."

"Certainly suits me," said the sheriff of Eureka, "if it suits Wells Fargo. I wasn't exactly partial to traveling alone with you boys, anyhow."

"Satisfactory to me," nodded Jim Hume, "until I run across your kind of boot-marks again."

"Then," summed up Mr. Neagle while Small and McDonald were taking Stewart's check and Wells Fargo's written absolution, "it looks like everybody's satisfied. There's some more of the boys outside that want to do business with Wells Fargo on the same basis. Meanwhile there's a custom in these parts that when everybody's satisfied, everybody joins in a certain little ceremony."

Nobody consulted Messenger Jim Miller.

# 19

# RICH MAN'S TOGA, POOR MAN'S SHIRT

It was an angry J. P. Jones who reached San Francisco late in October. Quartered at the Lick House within earshot of the stock markets, he pressed on to his new Mohammedan baths in Dupont Street. The stains of travel were boiled out quickly, but not even hammam hosings could quench that internal heat in a hurry. Jones had discovered that he had been let down by old allies—let down on two counts.

Ben Peart and J. D. Fry, understood agents of his close friend and past associate Alvinza Hayward, had neglected to keep him abreast of rousing events on the Comstock. This was plain faithlessness. Meanwhile their hands had been busy picking up worthless claims at Panamint and incorporating them for dizzy millions. It was obvious that the claims were being got ready for fancy stock-board acrobatics within the general beam of the J. P. Jones spotlight.

Peart was business agent for Hayward's innumerable projects. Fry was Hayward's broker. Neither would dare to affront the Senator without the nod of their principal. All this, in Jones' judgment, added up to but one conclusion. The baleful influence of Silver, which turned men's souls to iron, had entered the being of the man he trusted most and worked its black alchemy.

Between the two opponents in the shaping brawl lay half

a lifetime of successful adventuring together, begun in a day when Hayward, living in a cabin on the Mother Lode, had come to Jones' shack and struck him for a grubstake. Jones, good-naturedly helping his neighbor, had promptly found himself rewarded with an interest in a real vein which Hayward discovered. For a time the New Amador was the richest mine in California. When that million and a quarter in yellow quartz slipped out of Jones' clasp, Hayward sent him over the mountains to see what he could do with that certain run-down hole-in-the-ground on the Comstock. Jones proved that, given a Crown Point, he could do a great deal with it. All this was a story of robust and successful camaraderie, of faring together into reckless projects and places, and it had created one of the era's Damon and Pythian pairs.

And now, treachery.

Off came the broadcloth and linen of the thoroughly wroth Senator. He was stripped to his muscular buff and left by the attendants to perspire and vex. A stewing Hercules, grown pink and portly with honors and good living, there he sat. The goddess of the pale and mischievous metal may very properly have hovered to review him there—he was after all her darling. After a good steeping he was haled to the shampooing room, where his hammam mechanics fell on their boss with horsehair and cactus brushes. The ensuing spray of cold water did knock some of his dudgeon out of him. Toweled down, turbaned and enveloped in ruddy glow and a sheet, he was on his way back to a calmer state when they deposited him in the cooling room.

At this point the goddess of silver may well have skipped away, for her prankish work was done and she had plenty of other souls to torment. She left her sheeted favorite in a mosque-like chamber that had another occupant, similarly swathed.

It was Alvinza Hayward.

Hayward had lately been to China for his health. Still searching for health, he was now trying Senator Jones' public baths. So far he had found them salubrious, but he had not counted on meeting their owner.

It is difficult for personalities that regularly confront the world in boiled shirts, broadcloth, suitable trousers and heavy watch chains to produce quite the right effect on each other when they unexpectedly collide in the nude. The steam of the bathhouse gets between us and the ensuing situation. It must have been a mighty struggle between the Jones sense of humor and the Jones sense of indignation. In this case, indignation won. The tussle moved from the cooling room out into the rialto.

Hayward's wealth, like Jones', was still invested heavily in Crown Point. To drive his old backer out of that property was now the Senator's consuming purpose. At the moment, the great mine was listed at $55 a share. As soon as he could get back into his harness, Jones began to sell. He delivered solid blows in the market-place and waited grimly for the stock to sink. To his surprise, it stayed up. Jones hurled more Crown Point after the first lot. It was absorbed without incident.

This should have warned the angry man, but Jones' usual bump of logic was not on guard. He flung into the fray every share he had. When that failed of the desired effect he went twelve thousand shares short and then stood aside to watch, satisfied that he had smashed the bottom of the mine right down through the earth. Crown Point's price shook slightly, rallied, and swept upward to 67.

When Bill Stewart came down from Panamint in December he found Jones still putting packs to a headache and ruefully filling his shorts at any price demanded, convinced that every share he bought came from Alvinza Hayward and that his beloved old glory hole on the Comstock must have struck a new vein without his knowing it.

The row between Jones and Hayward, important if these had been any other times, turned out to be mere sideshow to a vaster drama. It was scarcely noticed by the populace that, a year before, would have been cast to the pit or lifted to perihelion by such strife between giants.

For the hour of tremendous events now was striking on the Comstock. Jim Fair, chipping away under Virginia City, had completely blocked out the south and west faces of his huge ore lump, the biggest ever found to that date within the boundaries of the United States. How far east and north would it run? The "big bulge" in the silver vein, its size and probable direction, was the only possible topic on every tongue.

Philip Deidesheimer, an accepted expert, went down in the depths of Consolidated Virginia and came up blinking. He estimated what he had seen, cut the figure in half for luck, halved it again for safety and pronounced that the metal in sight was worth at least a billion and a half. He was later supported in this opinion by a Director of the Mint. The lump extended from Con Virginia into the adjoining California and perhaps, at unheard-of depths, into historic Ophir. Now, with control of California and Ophir mines the prize, and Fair, Mackay, Flood and O'Brien meeting them blow for blow, Sharon and Ralston's ring were in colossal combat with the upstarts to wrest control of the Comstock's northern arm.

On the fifth of December, San Francisco's financial streets were black with people. Millionaires and sand-shovelers, businessmen and gamblers—waiters, bankers, Chinese laundrymen, butcher boys, cab-drivers—all were caught once more in the speculative whirl that was as characteristic of the town as the sea fogs blowing in through its harbor gate. Insiders looting outsiders; banks lending lavishly; clerks purloining from their employers and women selling virtue for tips—all was a mad, exaggerated repetition of the days of '71–'72, when San Francisco and Virginia City had spun in a giddy reel about Crown Point and Belcher.

This was not mere local extravaganza. The little city by the Pacific was, in these gorgeous Comstockian days, a stock board for the world. Chicago, New York, London, Berlin, Amsterdam stared wonderingly and bought in. And the gay city was as dark-hearted as it was glamorous. Every artifice, every form of chicane, every ugly shape of greed was present in some velvet or corduroy covering. It was common knowledge that the press was bought up or bribed. Commercial editors, assigned the future profits on shares which they did not own, fanned the bonfire with prophetic diagrams of the underground workings, glowing interviews with the superintendents on the ground, and lists of men already made wealthy beyond dreams. It was whispered that an editor of the tolerant *Alta* had already made $100,000 on a $300 investment for doing his part in carrying the market upward.

"The people are crazy," reported an eye-witness to the *Union* at Sacramento. "Yesterday and today, when California was jumping $10 at a bid, the frenzy of the crowd reached its height. The rush to get in was wild, enthusiastic and reckless. Brokers were almost beside themselves, while men and women rushed frantically about, issuing orders and eagerly scanning the lists and straining to catch the bids on the street."

Hour by hour the fortunes of battle changed. Lucky Baldwin emerged for an instant in control of Ophir. Sharon knocked him out of it. Sharon and Ralston seemed temporarily triumphant in the California. Flood and O'Brien hurled them out of that. Up, up soared California, Con Virginia, Ophir, Savage, Mexican. Everybody that could get a toe-hold in the shape of a hundredth of a "foot" on any 1080-foot claim was rich and getting richer. All was pandemonium. All was intoxicating, heavenly bedlam. "We have bonanza on the brain," editorialized the *Chronicle,* "hence we can undertake to write no article without putting bonanza in it. All other occupations are neglected for the bonanza. The grass withers and the grain dies; the plow stands still in the furrow: we are threatened with a dry season, a cattle famine and short crops. All this is unimportant, for the bonanza widens, deepens and grows richer, and stocks go up."

A temporary reaction caught Charles G. Meyer, who had come down from Panamint and paused to play the Comstock market. A man identified as the Surprise Valley assayer was fished from the bay a suicide on the morning of December 5th. Colin A. Spear, Senator Jones' private secretary, was another who joined too fiercely in the dance. He argued with a private citizen about the Comstock and received a bullet in the forehead.

When Stewart reached this silver-mad seaport he found his Panamint and senatorial colleague making his headquarters in the parlor of the Bank of California.

Gaming with millions makes strange playfellows. Though he hated Sharon as a barefooted man hates rattlesnakes, the genial John P. Jones now was paired with that cold genius.

A man of little magnetism for his fellow men but of much magnetism for wealth, William Sharon had long held dominance over Comstock affairs. Financed by the great bank of which Ralston was manager at San Francisco and he its Virginia City representative, he held mastery over Virginia's mines, mills, water, its local railroad, and for the moment its leading newspaper. Just now, Sharon wanted the north end of the Comstock as he wanted only one other thing in the world, which was to succeed the abdicating Bill Stewart as Nevada's senator. Jones in his turn wanted revenge on Alvinza Hayward and he wanted back the south end of the lode with his old Crown Point holdings. Staunchly the ill-mated pair now stood with each other, though dynasties were shaking.

To Bill Stewart's report on Panamint matters the junior Senator listened with half-interest. He also accepted from Colonel Baker the deeds to his new San Vicente Rancho at Santa Monica; and he heard F. P. F. Temple, up from Los Angeles, tell him that work on his Los Angeles & Independence Railroad must be rushed or the Southern Pacific would soon be up to the mouth of Surprise Canyon with its locomotives. The deeds he tossed over to a secretary and the southern banker he sent home with assurances that work on the L. A. & I. would start at once. Then he turned back to Sharon and Ralston. Now, if Con Virginia could be depressed just a little and the public frightened into letting go of adjacent holdings. . . .

A stagecoach lurched down Surprise Canyon from Panamint in the drenching rain of New Year's Day. It bore Harry Jones and Editor Harris among its passengers, and a freight wagon followed bearing Harry Jones' baggage in the shape of two big chunks of ore.

One lump was from a point thirty feet inside the Jacobs' Wonder tunnel. It weighed 782 pounds and was radiant with coppery blues and greens and shot with silver spangles. Its companion, a piece from the Hemlock, was smaller but still more radiantly charged.

A few days later the big lump, which assayed $500 to the ton, was gracing a sidewalk in front of a California Street brokerage office and the smaller piece, a sample of $1,000 ore, was flashing similarly from behind a window.

But the city refused to look. On and up roared Con Virginia, California, Ophir. At the turn of the year the total market price for these three Comstock performers equalled the assessed value of all the real estate in San Francisco, and no top was in sight. Sharon, who had turned from buying Ophir to buying legislators, was at the moment moving up and down Nevada urging "the boys" to snatch all the Ophir they could get but "not to hang on beyond 300." Recently it had been $60 a share. Now it was $225. The tip from the moneyed toga-seeker, aided by free alcohol, hurled it up some more. Tomorrow it opened at 270.

Con Virginia, $160 a share not long ago, now was $705. California, lately $89, stood at $790. At this price you could put in $60,000 and it bought you just ten inches along the Washoe fissure. But every clerk and waitress who held a knife-blade's thickness of ownership saw himself a new Monte Cristo who shortly would sit with Jones, Sharon, Ralston, Baldwin, Mills and Hayward, or perhaps on the cloud-wreathed summit with Fair, Mackay, Flood and O'Brien.

In Chicago the *Inter-Ocean* marveled: "No city upon this continent can show more men of solid wealth than San Francisco. Mines of fabulous possibilities pour their dividends. . . . Many of her citizens could sell out at a month's notice for $5,000,000 each. Palaces have risen from silver bricks, and the proudest buildings in the city owe their origin to ores and bullion."

"Worth $5,000,000!" the San Francisco *Chronicle* scoffed. "These are only our 'well-to-do' citizens, men of 'comfortable' incomes—our middle class. . . . In a small interior village like Chicago a man worth a million is esteemed wealthy. Not so in the grand commercial emporium of the Pacific. We do not call a person wealthy unless he advances beyond the tens of millions. Next year we will speak of those who possess hundreds of millions."

Into this hectic town, with its bay-windowed Palace Hotel mounting story by story skyward and its Comstock shares turning everybody into plutocrats, Harry Jones wandered to order his 120-horsepower engine and milling machinery for Panamint and Editor Harris to buy his new press and types. Jones and Stewart had departed. Fortunately for his feelings, the loser in the rousing side-scrap with Hayward had found fragrant balsam. Just after New Year's Day, in a home overlooking the Golden Gate, delightful Georgina Sullivan had become his bride, and no racket in the brokerage district could disturb J. P. Jones' idyl across the bay at rose-garlanded Tubbs Hotel, or the subsequent honeymoon in a palace car.

Congress was in session, but the two Nevada solons for the moment got no farther toward Washington than the eastern side of the Sierras. The fascination of the Bonanza was irresistible. Jones left his bride sidetracked in her car at Reno while he hastened to peer down into the familiar shafts. Stewart followed into the deep vaults a few days after.

Virginia City had grown since they last strode its hill-clinging streets. The town was now a warren of 25,000 people, one third whom were eternally underground toiling in heat and steam under tallow drips. Reverberations deep below were constantly rattling buildings and glassware at the surface. Public places were crowded, beds were working in triple shifts, new streets were scarring the terraced mountain, structures of brick were rising. There was still any game you wanted, with revolver and bowie knife refereeing, and the town joyfully proclaimed that it had more mistresses than wives, but it had always loved to wear its worst side outward and the straining eye could note numerous schools and churches. The cage under the Bonanza's hoisting works was skipping up and down two thousand feet in three minutes for the benefit of visitors who wanted to guess which side of a billion the great find totaled.

Nevada's only other congressman, Representative Kendall, must have found it as hard as the Senators to break away, for a measure was introduced in the Carson legislature sharply resolving "That the Hon. William Stewart and J. P. Jones be instructed, and Hon. W. Kendall requested, to proceed to the national capitol without delay, for the purpose of attending to their duties."

A few hours thereafter, votes were counted for Stewart's successor to take office with the March session, and Sharon, said to have distributed $800,000 in cash, was announced elected amid blazing bonfires and much synthetic rejoicing.

A terrific snowstorm hit the Washoes on January 15th, burying streets and putting the telegraph out of commission. Swiftly patched, the line was clacketing again—casting news, opinions, prophecies down to San Francisco and flashing back the ever-mounting stock-board quotations. In the week following, the silver-crazed communities on Mount Davidson had a real earthquake. Timbers swayed, joists groaned, brick façades ground and crackled. But the great mines held the moving rock apart with their stout square-timberings.

Three days later came shock of another sort.

For then it happened, with a catastrophic bang. Con Virginia looped from 715 down to 540. California dove from 780 to 370. Ophir, that first and oldest love, collapsed from 315 to a fraction of that sum, and Sharon's "don't hold on beyond 300, boys," became a shooting jest.

Then the story came out. Sharon, who had cheerfully paid three-quarters of a million to be tailored for a toga, had been shorting Ophir while urging his legislative dupes to buy. When he pulled the plug and the market price of that mighty mine dropped from $34,000,000 to $5,000,000, he coolly filled his shorts and got his political outlay back with a handsome profit. The legislators had delivered to him the toga plus their shirts.

This coup over, the stocks of the Lode set to their upward climb again. Brightest fireworks and blackest night were yet to come.

Against the flaming aurora borealis of the Comstock the two bright chunks brought down from the Panamints by Harry Jones glowed palely in their California Street brokerage setting.

# 20

# THE FIGHT FOR THE PASS

WHEN Senator Jones finally turned eastward, he left James V. Crawford fully commissioned to lay out the L. A. & I. on something more than paper.

There was need for hurry. Southern California was seething with railroad promotion and intrigue. Stanford, Huntington, Crocker and Hopkins, their Central Pacific between Oakland and Ogden triumphantly completed, had a government land-grant for their proposed new Southern Pacific down California to Los Angeles and thence indefinitely eastward. Jones had his dream of an overland artery slanting up from southern California to Salt Lake. Meanwhile overland immigration was pouring in by the Central Pacific at the north, and the southwest had virtually no rails whatever save a fragment of Southern Pacific running from Los Angeles a few miles west to Wilmington's tidewater. All other public transport was by stage or ship.

Crawford was a rising engineer with the Texas & Pacific, a paper route from the Mississippi to San Diego, when Jones bought him over and assigned to him the dual task of tying Panamint with the Pacific and Los Angeles with the nation. It was a commission to make any young man's heart leap. Crawford's did. He promptly took a bride.

Arrayed against him were the most formidable railroad builders of the day. Against him were formidable mountains, illimitable deserts. Against him was time. But Jones, the Silver Dustman, was at his back. Crawford left his bride under the orange trees and set to work. The initial problem was to route a way for rails up and over the San Gabriel and San Bernadino ranges to the Mohave plateau—that land in the sky studded with metal-charged knolls and gouged with salty valleys and dry lakebeds.

It was still summer of '74 when Crawford set out in a buggy with some sketch pads and toured from Shoo Fly Landing to San Bernardino, drove up a cleft between the two ranges that came together behind that village, and continued on by the world aloft to Panamint and thence to Independence. Return was made by Nadeau's freighting path down Soledad Pass. The reconnaissance convinced him that his outbound way, through Cajón Pass back of San Bernardino, was the most feasible corridor for iron rails. He further concluded that if iron rails couldn't be had he would lay wooden ones.

But by fall, with Jones approving and Panamint booming, there was no more thought of a wooden railroad. Early October saw the engineer—"Colonel" Crawford now—again in the field. He was staking out the line. From Shoo Fly Landing through Los Angeles to the Cajón the region was a bit vacant and dusty, but level as a ballroom floor. Up the pass went twenty curving miles, with an 1800-foot tunnel in prospect through soft red sandstone, and thence the way led onward over rolling country to a cottonwood thicket marking a crossing of the Mohave River—the Barstow of today.

Here the road would split, its main line ultimately to run around Death Valley's southern buttresses and up to a con-

tact with Utah Central and Union Pacific. That would put
Los Angeles on a transcontinental fork.

But the immediate job of construction would be a spur of
this, a narrow gauge paralleling Death Valley on the west and
tying the mines of the Panamints and Cosos with the sea.
Passing the mouth of Surprise Canyon, this branch would
receive its ore output and presumably its passengers by a
hilarious wire tramway. Ore that was now being transported
by mules, at $120 a ton, would be carried at $10 a ton by
Crawford's iron horse.

Local enthusiasm at Panamint pledged him a thousand tons
a day at the latter figure, and for supplementary freight the
Searles boys over in the Slate Range pointed to their deposit
of natural borax which they judged to be eighteen miles long,
eight miles wide and four feet thick—enough to keep ten
thousand men busy digging for fifty years, provided that
much borax was what the world wanted.

"The money being ready and the people men of action, who
have millions upon millions lying idle in Panamint waiting for
the road, there will be no time lost, and within twelve months
the shriek of the locomotive will be heard at the mouth of
Surprise Canyon," hailed the San Francisco *Post*—a Jones-
owned organ.

On October 13th the stake-driving started. On November 19th Crawford reported:

"We are camped on the summit of the Cajón, about 31
miles north of San Bernardino. It is very cold. Snow among
the pines reaches down close to our camp. Bears are numerous, and frequently interrupt the surveying."

But more than bears were sharpening their claws.

From the north the Stanfords, Huntingtons and Crockers
were watching. Their Southern Pacific tracks had been laid

a short distance southward from the Bay and now ended abruptly in the Coast Range wilderness. Certain other rails of theirs wandered up the great central valley of the San Joaquín, ending at no particular place.

Plainly these shaggy railroad kings were feeling their way; into the south their rails must go but they were undecided precisely where or whither. Matters were in abeyance anyhow while they tunneled the Tehachapi barrier. But now beyond Bakersfield, beyond the Tehachapi Mountains which horseshoed around the southern end of the great California valley, beyond the Mohave sandwastes aloft, were these new developments. The Panamint Mountains, which few men had ever heard of, were disgorging ore and promising to disgorge more, and ore was freight. Moreover, there was J. P. Jones, whom they little loved—who had tried to put a tax on their 7,000,000 Nevada acres—planning a railroad of his own that would not only feed on this freight and that of all Inyo, but so nourished would grow until it reached clear to Ogden and split overland traffic in the middle, diverting a full half of it over the Senator's proposed private tracks to Los Angeles.

The Stanford-Huntington group knew what to do with a challenge. Battle was their diet. To seize all passes and block the new line became business of the first order. Their rails reaching south into the hills below San Francisco Bay were left to terminate in a pasture. Straight up the interior San Joaquín Valley marched those other tracks, uncertain of their destination no longer, while in the horseshoe of the Tehachapis a yellow army, already hacking at the cliffs, was galvanized to redoubled action. To Washington rushed D. D. Colton, nominal president of the Southern Pacific, to demand before the House Committee on Public Lands an amendment to his land-grant that would give him the Cajón Pass and specifically "a new branch road affording shortened communication with the Panamint mines."

Los Angeles was aghast. The southern town with its population of a few thousand and ambitions for 1,000,000 had taken the lead in supplying the Stanford-Huntington crowd with $600,000 on the latter's promise to bring the Southern Pacific straight their way. Its diversion along the Mohave plateau and down the Cajón, thence bending east, would lead Los Angeles far off the main track. Southerners started a door-to-door canvas for funds for Senator Jones' road in retaliation; while Jones himself, now in Washington, waited grimly to give Colton's plan its coup de grâce when it should come before Bill Stewart's Senate Committee on Railroads.

But the struggle in the field had already taken place, and the victory had been won.

By the first of December Crawford had completed tracing and staking his route. And in December, Southern Pacific's hurrying rails were at Bakersfield. Shortly thereafter the latter were at the foot of Tehachapi's granite canyons, where for months the walls had been echoing to the crash of giant powder. Now the rising roadbed was looping and doubling on itself like a closely pursued hare around ever-mounting curves and through seventeen half-driven tunnels. At the foot of this pass another army of hundreds of coolies was laboring to create a sumptuous new wagon road reaching up into Inyo. The strategy of this fine new wagon road promptly became apparent. Remi Nadeau diverted his Cerro Gordo teaming to it, abandoning the long, hard 22-day haul down the old Soledad trail. In a twinkling the traffic that had done more to build up Los Angeles than any other single item had changed direction and was speeding for San Francisco Bay.

The Cerro Gordo freightmaster then turned on Meyerstein's sand-wallowing freight line and with a deep slash in his $120-a-ton rate put him out of business. Panamint's freights too went the northern way. It was a stunner for San Bernardino, which had been living in paradise. "Join hands with our natural allies [Jones and the L. A. & I.] and carry that Narrow Gauge through the Cajón Pass at a gallop. Time is everything, the Southern Pacific operates against us," cried the San Bernardino *Guardian*.

On December 11th, a party of Southern Pacific engineers passed through Bakersfield. On January 8th they were in the shadowy Cajón, setting up their surveying instruments.

Crawford was ready. An hour before Huntington & Co.'s transit men arrived, thirty-three brawny Jonesmen were at work under torrential rains felling trees and setting blasts in the echoing canyon. Storm waters, diverted from their immemorial course, roared and foamed around strange new fills, hasty trestle-footings. With foresight Crawford set other squads to putting up buildings against the bleak weather that was impending—a precaution which his rivals neglected.

Seeing the determination of the Crawford camp, the Southern Pacific forces sent a courier skimming over the plateau country for reinforcements from the grading camps in the Tehachapi. Chinese laborers were quick-hauled across the uplands. They found the Cajón occupied by a hundred of their countrymen under Jones & Co.'s banner—pigtailed, yellow-bronze athletes, many of whom had learned their road-grading trade under Charley Crocker himself in the Central Pacific's conquest of the Sierras. At every salient Crawford held the ground with his carts and his Chinese infantry.

Sheeted deluges now began falling. Newt Noble, coming down from Panamint with a long caravan and 52,000 pounds of sacked ore, reached San Bernardino after fifteen successive days of battling with the rainstorms. With him came a section of the Chinese driven out by Small and McDonald; in Cajón Pass these found refuge under the tents of their countrymen and provided Crawford with welcome reinforcements.

Black as winter skies had been, on the sixteenth of January they went blacker. A regular Forty-niner downpour set in and continued for three days, growing harder each hour. The twenty-sixth launched a genuine blizzard, the roughest snow-hurricane seen in many a day. Strong winds with the force of "Washoe zephyrs" ripped down from the north, driving the snow in blinding clouds. That ended the ambitions of the Southern Pacific party to snatch the pass. Just barely forestalled by its better-domiciled defenders at every point of physical possession, but forestalled nevertheless, they fell back on the tighter, snugger camp below the distant Tehachapis, their retreat under the lashes of the highland gale becoming something of a rout.

Enthusiasm for the feat of the Jonesmen soared to the skies.

"Beyond the line of those majestic mountains, little more than fifty miles away," cried an orator at a boosters' celebration in Los Angeles Valley, "you come upon a region as barren as that which fringes the Dead Sea, as desolate as the black rock upon which Aden stands; a land of sagebrush and cactus, where not even a leek or a potato can be raised. . . . But locked up in those almost impenetrable recesses lies immense, incomputable wealth. By the genius of John P. Jones, and our own co-operation, those treasures are about to be unlocked. . . ."

I have myself been a miner for fifteen years, in California, Nevada, Idaho and Oregon [continued orator Frank P. Ganahl.] I have been to Panamint and I am prepared to say that any one of the three mining sections there will yield one thousand tons of ore a day. . . . One thousand tons of ore represent ten thousand miners with stomachs. They must be fed. They need your pork, barley, corn, wine, brandy and fruits. . . . Through the dash and engineering pluck of your chief engineer, Mr. Crawford, you have secured the Cajón Pass from the Southern Pacific. Over that pass lies your market. Two hundred miles brings you to Panamint, and 480 to a junction with the Union Pacific at Ogden, securing you a continental railway. . . . All John P. Jones and his associates ask is that we shall raise in the three counties of Los Angeles, San Bernardino and Inyo, $300,000. If we raise that sum, those gentlemen will put in another $300,000 and bonds can be issued for the rest. . . . The profits will be immense. The prodigious returns from Sharon's Truckee & Virginia Railroad show what a mining railway can be made to pay.

"The whistle of the locomotive will soon be heard," exulted local journals, "and oranges picked in San Bernardino in the morning will be served in Panamint for supper."

THE GALE which drove Stanford, Huntington & Co.'s invaders out of the Cajón put a stop to all wagon navigation on both sides of the Sierras and piled up snow so deep in Surprise Canyon that teamsters had to abandon their wagons. The harsh weather made work impractical in the mines and there was a general layoff ordered, with abrupt effect on the pleasant chink of the circulating medium. With too many people, little food and no work, a state of fret developed, breaking out in a rash of shootings and unpleasantness. The best cure for it all was action and the best kind of action was hunting for lost mines and new, and prospectors who found hibernating not to their liking hit the trails in all directions.

One of the first of several brilliant strikes was made on the side of Telescope Peak. Joseph Nossano and his brother uncovered ledges averaging $919 to the ton in silver-lead, and word of that one sent the burro-tailers flocking. "I am of the opinion," assured one sage of the desert to the Independence newspaper, "that the Maria in Wild Rose is the lost mine known as the Gun Sight." In a week it was forgotten.

A dozen strikes in other directions set prospectors energetically to loading up their jacks. One party made up the canyon back of Panamint, over the high wall, and down the slopes beyond, resolved to track down once and for all the will-o'-the-wisp reported by Jacob Breyfogle years before. Off across Death Valley they tramped and up many a wash and notch on the opposite side, until out on the edge of the Amargosa sandplain they decided they had found it. The word sped back to Panamint. "This celebrated ledge, supposed to surpass the great modern bonanza in wealth, has thrown Panamint into a state of the most intense excitement," chronicled the Panamint *News*. The excitement lasted about five minutes.

James Bruce was one who joined the legion of fareforths. Abandoning temporarily the faro box and the undertaking business, he took the aboriginal known as Indian George—the same that had nearly led prospector S. G. George to Panamint's cliffs in 1860—and was conducted by that brave over the stiff ridge south of camp. Bruce was armed and his hand was quick, but his motto in life was to keep all play in front of him and he now said: "Captain, this is Indian country. You know it, I don't. You go first."

They camped that night in a barren gorge neighboring Surprise Canyon, Bruce sleeping on his arms and waking in the morning to find himself bedded on a ledge of such opulent embrace that he monumented it at once, called it the "Juno," and broke off forty pieces. As they returned over the divide and the squat, brush-laden roofs of Panamint town came in view far below, Indian George halted and pointed down in his turn.

"Ugh," he said, "white man's country. White man knows it. Indian don't. White man go first."

"Fair enough," Bruce had to acknowledge, and swung on down to camp, Indian George following under a sack of ore that later assayed $80 to $420 to the ton.

Still the reports flowed in, the *News* announcing: "Prospectors who have come in within the last few days, bring with them exceedingly rich rock, from all sections—east, west, north and south. They are highly elated, and say that the country hasn't been prospected at all yet."

Not that the future of the region was gorgeous to every eye. Surprise Valley was yielding high-grade at a profit even when transported for a couple of hundred miles by packmule and shipped over two oceans. But the Great American Desert was studded with dead camps once hailed as "second Comstocks." This high-grade might not last. Cheaper handling would then be imperative. Until mill chimneys were belching in Surprise Valley and iron rails ran to the sea, most of the profit would be claimed by mule skinners and ship masters.

This conclusion forced a heavy laying-off of men in January, giving the camp its first setback.

When Kiv Phillips came home to his cabin one night and

found that the meanest man in several deserts had made off with his only spare pair of pants, he decided that the place had seen enough of him. Three weeks later he was back in Eureka, asserting that "if they put the whole town of Panamint in a blanket and shook it for an hour, nothing of value would fall out, there not being a two-dollar-and-a-half piece in the whole damn place."

Down the canyon and back to the central Nevada town also trudged old Uncle "Billy Bedamned" Wolsesberger, whacking his little burro and retracing the whole four hundred miles on foot. He brought word that none of his fellow-Eurekans had made their fortunes up there; that the region about Panamint was overrun with guerrillas and road agents; that he had formerly regarded Eureka as a little tough, but was thankful now to get back to a place where a man's life was moderately safe. The day he left Panamint, he told old friends, the stage had been stopped and all its passengers touched for their belongings.

Spring and the Seventeenth of March came simultaneously; the first with a sudden bursting into bloom of blue sage and yellow cassia on the mesas and washes below, and the latter "celebrated here," dilated the *News*, "by the firing of a green flag, a large blazing anvil and the hanging of bonfires on several doors. In the evening nearly all of the saloons made speeches in front of the speakers, and a good time was had by everybody."

By this time the purpose of the big chunks of silver-charged rock brought down to San Francisco had become apparent. Twenty thousand shares of Jones & Stewart's Wonder Consolidated and Wyoming Consolidated properties, the choice lodes on both sides of Surprise Valley, were being offered to the public at $15 a share.

Had the Silver Dustman decided that those heights, after all, contained but superficial glitter? Or was the moneyed Senator, following his bout with Hayward, beginning to see the bottom of his sack?

Despite the leaps and tumbles of the Comstock, which held all men's eyes, and the unhappy walking ghost of Little Emma, enough of the Panamint stock was sold to cause the dispatch from San Francisco via Wilmington of the chief iron elements for the Surprise Company's tall new mill. Hailed the San Francisco *Alta:* "The whole will be in running order within the next sixty days. With such a prospect, and with the great richness of the ores which are now being developed in these Panamint mines, . . . the progress and quick development of this new district will shortly be such as was witnessed in the early days of Washoe."

Meanwhile, on a neighbor hill, a very definite rival already sat combing out her hair.

"The purpose of the big chunks had become apparent."

# 21

## THE RIVAL

THE Circe who perched in the sunlight thirty airline miles away was a siren of Panamint's own summoning.

William D. Brown, who first called her to public attention, was a cut above the burro-and-frying-pan prospector. He was a mineralogist, sometimes known as Professor Brown. Veteran appraiser of rocks and holes, he based a roving existence on an address at San Francisco and experted treasure strikes and rumors of strikes for promoter-clients. Bill Brown knew the California-Nevada border and seven years before had turned in a report, written in a fine copperplate hand, on the prospects of Cerro Gordo when that galena-charged hill was the latest topic.

At that time, Brown had wondered if the Cerro Gordo ore belt might not continue on with a long southeast swing. The surmise was put aside at the moment and had since been almost forgotten. When suddenly, this tumult in the Panamints. Might that outcrop not represent the other end of his suspected belt? And if riches lay at the extremities they might very well lie between.

Bill Brown had been reaching for fortune's silver girdle for so long that he was all packed up to follow whenever he heard her pattering feet. With his brother Bob he left San Francisco in October of '74, making as speedily as possible for the young camp under Telescope Peak.

From that boisterous settlement, where they tarried for a few days to acquire supplies and sift rumors, the two moved out with considerable round-aboutness. Professor Bill's theory, until proved wrong, was worth protecting. Old Coso in the folds two ranges west was his first objective.

That pioneer camp had dwindled in its dozen years of existence from a couple of hundred men to a bare handful. Its residue of Mexicans had been working up their gold quartz in rude arrastras when Jacobs, Kennedy and Bob Stewart had dropped in two Christmastides before. Since then, as for years preceding, nothing much had happened in Old Coso. A little gold, señores, and a little liquor. When, all unexpectedly, the goddess of Silver chose to show her face.

The pale hussy must have been tagging close to Bill Brown all the time. The visit of the Browns to Old Coso coincided with the return to town of one of its regular residents, and what this Mexican had with him caused his fellow-townsmen to drop their work and gather close. The Mexican's jack was laden with rock wholly alien to the neighborhood, for it was heavy with white metal—argentiferous galena.

Professor Bill joined the throng and handled the samples just brought in by Rafael Cuervo. As he weighed them in his palm he knew at last the thrill he had been seeking for many long years. There are moments when a man must study his control. Mostly base metal, he remarked as he handed the chunks back to their owner. Where did the excellent gentleman encounter them?

For answer the excellent gentleman gave the professor an enigmatic flash of white teeth under leathery cheeks and black moustaches. At a great distance, señor.

Was there more of it?

Sí, señor, a whole mountain, if truth must be told. A so full, a very full mountain, though escondida—the finder's secret.

Brown's blood was thumping in his ears. A mountain of this sort of stuff could mean only one thing. The Gunsight, nothing less. The real Gunsight, at last—and Bill Brown's. Though he must keep command of his face, for this was high-stakes poker.

Once more he steadied his voice. Would the black-moustached gentleman, the very distinguished hidalgo, care to indicate the direction from which he had just traveled?

The dusty Mexican, still beaming, obliged by waving his wrist in a wide arc to southward. "About two days, maybe four, in that way, señor."

Messrs. Brown thanked their informant politely and stayed overnight for appearances, though itching to be off. In the morning they left Old Coso, moving straight south until well out of sight of its flat valley, when their direction took a horseshoelike turn. Professor Brown knew his Hispano-Californians. That mountain lay directly opposite the way its locator had gestured.

As he rode, Brown studied every rock and knoll. The hills were composed mainly of lime to the west, porphyry to the east. The materials were tending toward a junction somewhere straight ahead; and that junction was what he wanted. The brothers pushed for a crest. It stuck up three thousand feet above the hovels of little Coso, now four miles behind them but seemingly not half a mile away.

From this treeless summit a wide sweep was revealed. Owens Lake, that unhappy dead sea, lay on their left—a mile down, twenty miles away. Springing beyond its western shore, first in uptilted sage, then in precipitous granite, the grand white wall of the Sierra Nevada lifted up, up, up, until it hung two and a half miles above all troughs and deserts. Straight north and above Owens Lake the furnaces of Cerro Gordo on their high pedestal stained the sky. The Argus Range on the travelers' right, rising to a crown slightly higher than their present vantage point, shut off the Panamints that held Surprise Valley, though nothing could obscure the blue-cowled head of Telescope Peak, overseer of all.

Closely Professor Brown studied his terrain, searching for the meeting-point of those lime and porphyry ridges, and concluded that they merged on a smooth bare mountain about eight miles onward. Thither the pair proceeded, moving with growing confidence, though forced to return that night to a spring near their lookout point—there was no water beyond.

In the morning the brothers were again on the trail. They were following domestic animal tracks now; their predecessor had not thought to cover those. They came to a long arroyo which ascended from Panamint Valley; its head was in a fold of the bare round mountain they had seen the day before; and at the top of this arroyo they came upon a hand-made pile of rocks—a prospector's notice of location.

Bill Brown drew from a crevice a freshly-scrawled paper. It was signed by Rafael Cuervo of Coso and the Browns knew that here, beyond much doubt, there would soon be a wild new town and district.

For some days the brothers busied themselves monumenting new prospects and chipping off outcroppings. Then, beside the greasewood fire of their dry bivouac, they might have been glimpsed applying final tests.

A pair of balances had been set up, their little pans dangling to receive platinum weights in one, a shaking of pulverized ore in the other. From each lot to be assayed, a portion of rock was meticulously weighed out corresponding to a 29.166 gram weight in the other pan, simulating "one assay ton." Brother Bob packed hot coals about the little fire-clay box known as an assayer's muffle and into this oven Bill Brown pushed his dishes of measured ore.

Occasionally he stirred with a piece of wire set in a wooden handle. After the charges had reached a red heat and gave off no more fumes or odor the professor removed them, let them cool, and transferred them to crucibles.

Like a Merlin working incantations he proceeded now with his chemist's brew: dropping into each charge its due of litharge, borax, charcoal, carbonate of soda and carbonate of potash; stirring, covering with common salt, and placing several lumps of borax on top of each. Once more Bob Brown had the coals very hot and now the crucibles with their puddings went into the furnace. When the contents were fluid the crucibles were tapped lightly, allowed to cool and then broken open. Buttons of white metal remained.

The next act in the intricate little drama going on without an audience in that high arroyo was to segregate the metals composing these buttons, a process known as "cupellation." When an alloy of lead and silver is heated sufficiently in the presence of bone ash, the lead melts, oxidizes, and is absorbed into the ash. What remains is "noble metal"— silver or gold or both, to be separated in turn by acid treatment. Each milligram of silver so produced from 29.166 grams of rock indicates the presence of one troy ounce in a 2,000-pound ton of ore.

Checking and rechecking, the Browns rose from their work more than satisfied. Fifty per cent lead and seventy ounces of silver to the ton—so the tests revealed. And over to the

# The Rival 229

county seat at Independence hurried the pair to record their claims.

Up, down and across the sunlands then, from screw-bean mesquite to ocatillo, from rabbit brush to ink-weed chattered the wasteland telegraph, of bright new ledges "twenty feet wide, extending a thousand, two thousand feet."

The word spread to Lone Pine, Panamint, Old Coso, Cerro Gordo. It raced for ever-widening horizons—to Pioche, Eureka, Austin, Carson.

Jack Wilson, nee Curran, one of the originals at Panamint, threw on his horse a thirteen-shot Henry rifle, a Murcott hammerless shotgun, and a Winchester forty-four-forty; put in his belt a couple of the largest-size breakdown-actioned products of Messrs. Smith & Wesson of Springfield, Massachusetts; and was off before his old camp knew he had breakfasted. He rode in on one Mexican who was admiring his own new claim, told him urgently that he didn't own it, and for law referred to the arsenal on his packmule. He named this seized mine the Defiance.

First arrivals pitched the usual crazy tents, burrowed into hillsides and piled up stones in front of dugouts. Their successors brought in cargoes of lumber by packmule. There was no water, no timber within miles. But there was silver.

"ANOTHER STRIKE—*Panamint Rivalled,*" headlined the Bakersfield *Californian* on December 17th. "By a private letter from one of the discoverers, we have news of another remarkable mineral discovery. . . . Assays have been made with the most gratifying results, the lowest return amounting to $700 per ton. . . . Great excitement prevails. . . . From the known extent of these mines and their indubitable character, we are justified in saying that everything points to a time not far distant, when they will rival Panamint, or even the Comstock itself."

On the road chanced to be Lucky Baldwin, that legendary plunger. Somewhere between Bakersfield and Panamint the sagebrush rumors caught up with him, and presently the *Bulletin* at San Francisco was chronicling of him and his party: "They bring specimens of ore. . . . The first, they think, indicates the richest in the world, and the latter richer still. They go to San Francisco to provide the way for opening the mines, and promise developments that will challenge the Comstock."

Men were still mounting the canyon that led to Panamint, but the new lode hastened more than a trickle of departures. Senators Stewart and Jones, while at San Francisco, were able to read in their *Altas* the morning after Christmas: "Latest advices from Coso confirm the reports of the richness and extent of the recent mineral discovery. A great many locations are already made, and prospectors are arriving daily. There are about 200 men now at the camp. . . . The proposed name of the town is Darwin"—honoring Dr. Darwin French, the Gunsight searcher of a decade and a half before.

January found the San Bernardino *Guardian* conceding the new camp to be "looming up as a rival to Panamint." March found Darwin so far along in civic importance that road-agents were making things lively in its vicinity. April brought to it the regular service of the Lone Pine–Panamint stage-coaches, the vehicles leaving Panamint in the morning and by means of a fifteen-mile detour depositing their passengers at Darwin in the evening.

By May, with the desert air heavy with the perfume of primroses, encelias and verbenas and travel on the southward-leading trail increasing, Virginia City bestowed the accolade of recognition upon the new contender. Said the *Enterprise*:

> There is considerable talk about the new district of Coso, situated a short distance from Panamint, and several gentlemen of this city contemplate starting for there in a few days. The report of the discovery of new diggings is to the prospector like the sound of the trumpet to a war horse; and tell an old forty-niner that a new camp has been found, and he commences rolling up his blanket forthwith. This morning we heard one of this genus expressing his intention to start at an early day. Said he: "I've been to every excitement from Gold Lake to South Mountain, and I've mined in every camp from 'Tuwollomy' to Fraser River. Since I came over on this side of the mountains I've traveled over every mountain and desert plumb to Death Valley. I haven't never got rich a-prospectin'; but I know there is good diggin's in the country, and I'm agoing to keep on a-huntin' 'em till the Old Man calls me to glory." The distance from here to Coso is over 250 miles, over a barren and desolate country most of the way; but that will not deter this or any other prospector from starting to go there "afoot and alone."

Early to go clattering down Surprise Canyon and over the sinks for the new rival camp was one-armed Pat Reddy. The buggy that had hoisted him to Panamint now lurched and rattled for the head of Darwin Wash, where Inyo county's legal light bought up Jack Wilson's nine points of armed possession in the Defiance mine, opened a pioneer tavern, and sent in haste for his brother Ned to come over from Panamint and operate both. And though litigation raged through the courts for several years, the Reddys hung on to the Defiance and in the end Pat made Jack Wilson's roughly won title stick.

Down from the Panamints reconnoitered Small & McDonald, and up the side of the bare peak in the Cosos so rapidly acquiring a brand-new town; but as the approaches to Darwin contained not half the tactical advantages of Surprise Canyon, they concluded to go back to the old and well-tried—in fact sorely tried—love.

Another pair of shrewd eyes had also begun to doubt the depth of the treasure behind Panamint's cloud-scoured walls. R. C. Jacobs, leasing his wonder mine and its mill to Dave Neagle, joined the migration to Darwin.

Promptly for the new field, and in a hurry, moved yet another, one Nick Perasich. He was a restauranteur who had not reached Panamint quite early enough. But he was determined to get in on the ground floor at Darwin. Perasich pushed on with celerity, keeping an eye out for horsemen, men in stagecoaches and men in buckboards.

Perasich had left behind him at Panamint several feuds and unpaid bills, and he had a feeling that however fast he traveled, someone might be moving on his trail still faster.

# 22

# THE TURBULENT CASE OF BARK ASHIM

THROUGH the icy discomforts of that winter, tempers at Panamint had flamed and sputtered like pine knots and Will Smith's judgments had been in frequent demand. That he decided all matters with easy indulgence was perhaps the only thing possible under the circumstances; the witnesses usually sided with the survivor and had been under the tables anyhow as soon as the shooting started. Inyo's only jail was at Independence and that shire town was at quite a distance. "Self-defense," Will Smith would rule, and that settled it.

Yet there was a mechanism for formal justice. The Hon. Theron Y. Reed served the 16th Judicial District of four large counties, one of which was Inyo. Judge Reed's circuit was 250 miles long, about 100 miles wide; he sat successively in Independence, Havilah on top of the Sierras, Millerton over on the edge of the California great valley, Markleeville far at the head of the eastern Sierra country, and Bridgeport in the vicinity of Bodie and Aurora.

This orbit swung him around the loftiest and perhaps wildest 2,000 square miles in the United States. When the judge rode circuit, he went heeled—a man never knew when he might meet a grizzly bear or a disaffected defendant; and when he mounted bench he kept at hand the same useful implements. Thus he is said to have settled one of the first objections propounded by Pat Reddy by cocking both barrels of a gun and resting it beside his chair. On another occasion a well-aimed tome brought the fine points of the law home to an attorney and convinced him of the accuracy of a ruling. But Judge Reed's firmness could go no farther, whereas the adeptness of Pat Reddy at getting his clients off the hook knew no limits whatever. Pat made his headquarters at Independence. And for some years there had been little use in taking Inyo killers to Independence to appear before its judge.

It is a genuine loss to frontier drama that Pat Reddy, a hoodlum of early Bodie who advanced by native wit to become one of the ablest attorneys of the silver belt, came on the scene about one decade too late and four counties too far south to match legal sword and mace with Bill Stewart.

For it was Reddy, more than any other single human influence, who was responsible for the peculiarity of law and order in Inyo and outlying places. Reddy simply scared, mesmerized or outwitted jurors until "not guilty" was as certain as salt in a bucket dipped into Owens Lake. Seventy homicides, according to contemporary count, already studded local history with only one conviction. And the one conviction did not appease the lovers of orderly justice any. For E. P. Welch, the killer whom Pat had failed with, had recently tired of awaiting execution, noticed his jail door open, and with calm impudence picked up his leg-ball and decamped.

Seventy killings. Sixty-nine official ignorings, dismissals or acquittals, and one escape. Into this volcano of rising popular indignation hurried the young Jew, Bark Ashim, intent on laying his hands on the man who had wronged him.

Nick Perasich, the quarry Ashim was after, with a fellow Dalmatian named Petrovich, had operated a restaurant and vegetable mart on the ground floor of a shack on Panamint's Main Street, and Nick in partnership with unamiable Cervantes, a Mexican, had also conducted a groggery in the cellar. The ground-floor kitchen was presided over by Constantine, a Greek. This international arrangement might have worked out harmoniously in pleasant springtime and an era of active business; but in lean winter, with half the camp laid off and file-like winds blowing up the canyon, something had to give.

Perasich had never been known to shoot anybody, but he strutted a good deal and was known as a quarrelsome, browbeating fellow. When the presence of his Mexican partner became unendurable he broke up a wine bottle into very fine bits and offered Constantine a bonus if he would bake this brittle flour into a special cake for Cervantes. Constantine objected, and was fired; demanded his back wages, and was given a fictitious order on someone over in Darwin; journeyed there, and failed to collect.

It was while he was returning from this effort that Constantine encountered his employer on the road in the Argus Mountains. Once more he demanded his wages. This so exasperated the departing Nick that he sought to beat reason into the chef's head with a revolver handle.

It was a sad reward for all those compotes and ragouts. More dead than alive, Constantine crawled down, across and up to Panamint and there swore out a warrant for his employer's arrest. But Perasich had shaken the snows of Panamint from his feet and no peace officer could be found who desired him enough to go after him.

Perasich pushed on to Darwin.

Among those who joined with Constantine in regretting his departure was Bark Ashim, who had reached Panamint in January after adventures on the way as previously related. Ashim's family and the mercantile house of that name were well and widely known in Carson, Eureka, Hamilton, and elsewhere in the silverlands. Before lifting the family banner in Panamint, young Bark had tried store-keeping at Pioche, where, on a charge of assault to commit murder, he had got himself arrested and fined a dollar—Pioche being then in the throes of a mild reform movement. Otherwise his record was not tempestuous. He was a cautious young business man doing the best he could for himself in a lusty environment.

While in Panamint he had seen numerous cases go to final judgment for little or no basic cause. Compared with the trivialities over which his neighbors had frequently fought, a sum say of forty-seven dollars and a half loomed bigger than Sentinel Peak at the head of the canyon, or Maturango against the sunset in the Argus Range opposite. Consequently when Bark found that Perasich had left camp for good, leaving unpaid his account for exactly forty-seven dollars and a half, he picked up a Whistler six-shooter, took along his clerk Sam Montgomery, and caught Billy Balch's outgoing stage.

It was the tenth day of blustery March when Ashim, dropping into a Darwin restaurant, saw Nick Perasich sitting at a table with his partner Petrovich. There were words, a taunting offer on Nick's part to pay "when he got back to Panamint," a bit of insistence on Ashim's part that he pay up now, and a gesture toward his holster on the part of Perasich. It was the eleventh of March when Ashim entered Sullivan & Milstadt's restaurant again and for the second time beheld his debtor. The affair was still in the hard-looks and hard-words stage, and might have advanced no farther, save that this time

Ashim was accompanied by an ally named Tom Carroll.

Carroll was one of the tossed-out ruffians of Pioche who had recently made his way to Darwin. There as a hopeful real estate purveyor he made it his custom to attach himself to visitors. In this capacity he had attached himself to Ashim the day before. Carroll felt sure that Ashim would see the possibilities of Darwin's town lots as soon as his mind was cleared of whatever was bothering him. To clear his client's mind of what was bothering it seemed therefore to Carroll to be the first step toward a stroke of real estate business. So when he entered the restaurant behind Ashim, and saw the black looks exchanged between his prospect and the Dalmatian in the corner, his hand slipped easily to his side pocket.

He was in this posture when Ashim strode to his antagonist and said: "Perasich, I want you to settle that account! You will not put me off as you did yesterday."

"I will settle that in Panamint," vowed Perasich.

"You have no business taking you back to Panamint and you will settle today," said Ashim.

Perasich sprang up. Thereupon things happened so fast that it took several witnesses several days to explain the sequence of the next five seconds.

John Sullivan, the cook, had emerged from his kitchen at the rear. He leaped for it again, bowling over an intervening screen.

A stride behind him came his waiter-partner, Milstadt. Ashim said in the trial afterward that he fired only at the baseboard, and this to scare Perasich who was coming at him. A bullet from his gun was later found embedded along the line of fire as his testimony indicated. His second shot, he said, was aimed at Perasich's gun hand. But Perasich reeled and slumped with two bullets in his chest; and Carroll, who evidently shot on principle at any bird on the wing, then let fly at the waiter—missing but speeding him up considerably. Not to be outdone in the general gunpowder symphony, Perasich's partner Petrovich also leaped up, got out a little pocket I.X.L. pistol, and in firing it valiantly shot off two fingers of his own left hand.

When the cacophony ended, Perasich was dying on the floor, ten or a dozen shots were embedded in the woodwork, Tom Carroll had fled never to be seen again, and Petrovich, Milstadt and Sullivan were racing with extraordinary agility for places elsewhere.

Darwin, though a very new camp, had already had an undue amount of bloodshed. It not only had drawn heavily on several tough towns for its residents, but here were Panaminters dragging their Panamint-bred quarrels over the sands from Surprise Canyon for settlement. Knowledge that Pat Reddy had snatched up Ashim's case almost before the guns stopped smoking made the matter a real issue.

Bark Ashim must be lynched.

But Pat Reddy had foreseen this too, and when the vigilance committee closed on Ned Reddy's dramshop, whither Ashim had been hustled, its intended victim had vanished. After a wild ride in Pat Reddy's useful buggy, Ashim was transferred to the keeping of Sheriff Moore at Independence.

It was at this juncture that E. P. Welch, Inyo's only convicted killer, elected to make that insolent get-away from the courthouse jail with his leg-ball gripped in his hands. Ashim was just unlucky enough to have mixed himself in a manslaughter case at the moment when the high gods were laughing at law and order as it was administered in Inyo. That settled it. What the populace of Darwin had failed to accomplish must be performed by public avengers at the county seat.

While Ashim was awaiting trial, the vigilantes banded. They came from Bishop Creek on the north, from Lone Pine on the south, from deserts east and from the mountains west. They were three days collecting. Scouts were thrown out and camp was made on the far side of the Owens River. On the third night they moved on the flimsy courthouse.

It was oddly dark and silent. The conspirators had expected to find it ablaze with light. The blackness could suggest only one thing—every window must contain a loaded shotgun or a sharpshooter with a rifle. It would be just like Pat Reddy to arrange matters in such fashion. First to admit a waning enthusiasm were a squad captain and his force of six. These deserted, and the remaining twenty-one adjourned to the greasewood on the far side of Independence and held a council. Their number was augmented by at least one other, who came up in the darkness and seemed to have all the passwords. It was vigorously moved and seconded to fire Reddy's house and barn, hang the barrister if he could be found, hang his clerk for keeping such bad company, and while the flames were drawing all eyes assail the jail and seize its prisoner.

To this general proposition a single voice dissented. It was the voice of the newcomer, raised in the outer darkness.

Would it really be sporting, questioned the voice, to hang a lawyer simply for doing the best he could for his clients? Besides, with Pat Reddy hanged, where would the vigilantes themselves look for aid when they got into trouble—as they often had in the past, and surely would again?

To shouts that Reddy had openly threatened the trial judge's life, thereby assuring a favorable outcome for Ashim, the voice reported that Judge Reed always rode circuit and sat at his bench armed with two heavy revolvers, so was presumably prepared to take his chance like any other man. As the logic was impressive and presented in Pat Reddy's most persuasive manner, backed up in the darkness—so the listeners suspected—by Pat's own gun, the vigilantes agreed to forego the pleasure of hanging him and melted away.

Reddy wasn't done. He promptly swore out a warrant for conspiracy against four of the vigilantes and published this ringing defiance:

"In consequence of the fact that I am engaged as counsel for B. Ashim, now in our county jail on a charge of murder, together with the heinous crimes of having acted as counsel for other persons heretofore accused and acquitted of crime in this county, I am informed that the Vigilance Committee of Bishop Creek, or rather a certain portion of that organization, in solemn counsel on last Friday night at their office in the sagebrush about one mile south of this town, determined *'to burn my property and hang me.'* To those cowardly wretches who engaged in this conspiracy I have this to say: That I will continue to practice my profession, will defend or prosecute whenever I see fit to do so, and I hereby hurl defiance in the teeth of all midnight assassins and marauders, and especially to those conspiring against me. True, you may murder me, but you cannot frighten or intimidate me."

Soon after, the body of the fugitive Welch was found on the edge of the desert, emaciated and gnawed by coyotes. Little was left but the iron ball. Justice of a sort had caught up with Pat Reddy's sole convicted client.

In the little Balkan village which had produced the Perasich brood, family ties are close and there is much vigor in the law of an eye for an eye, a tooth for a tooth. When Nicholas Perasich and his two brothers Elias and Peter sailed away for America they brought this ancient law with them. Elias and Peter had gone to Carson City in pursuit of fortune, Nicholas to Panamint. Ashim was bound over for trial in bonds of $15,000. In getting out of town he experienced a good deal of anguish, for Elias and Peter Perasich had come to town to have a hand in his leave-taking. The defendant got away only by the superiority of Pat Reddy's intelligence department and rapid use, for the second time, of Pat's buggy.

Six months later Ashim, who had been visiting in San Francisco, returned for trial. He traveled by rail to Carson City, having arranged to meet his mother and sisters at the latter place, with passage provided for in the stage from Carson southward.

The Ashims were former residents of the Nevada capital. The mother arrived first. She went to the Ormsby House. There she was recognized by a waiter, who communicated his knowledge to the two living brothers Perasich.

Bark reached Carson circumspectly. He went to the Ormsby House at 5 A.M., just before the stage pulled out. A friend of the family got word to him that Elias Perasich was standing behind the dining room door with a shotgun.

Bark left by the kitchen.

Following plans that had been laid for such a contingency, he pushed on down the road afoot. When the stage drew up before the hotel it took aboard the mother of the pedestrian; her two daughters; a spinstress; a matron and children; and two gentlemen in borax who were bound for Columbus, the Messrs. V. Ferbach and S. Pinchower.

When they found that their quarry had eluded their trap in the dining room, Elias Perasich and a supporter rushed to the porch and attempted to book passage in the filling coach.

Its driver, sniffing trouble, declined to accommodate them.

The avengers had dressed hurriedly. Joined by Pete Perasich, also garbed in haste, they set off at a smart trot down the highway with flapping tails and open shirts.

Ashim was half a mile ahead. He saw the foemen before they saw him, jumped into a cornfield, and hid in a ditch. The pursuers asked a boy, who approached driving some cows, whether he had seen a man passing that way. The boy had noticed no one. The pursuers ran on, turned back and began searching the ditch.

They were a dozen paces from the hidden man when the stage came abreast. With a flying leap, Ashim was out and on the swiftly moving step. Elias Perasich elevated his gun and sent a charge whizzing. Ashim fired one pistol shot in return, then made it in through the window, assisted by sisterly and motherly hands,—to open the door would have pushed him off the step. The six horses high-tailed it with a clatter. Another shot ripped through the rear of the coach close to Ashim's head as he was falling into the cushions and narrowly missed pinning the maiden lady's bonnet to the leatherwork. The Perasich party then took to the balls of its collective feet and made a dash after the stage, firing as they ran, with Ashim returning shot for shot out the coach side-window.

On the hurricane deck of the vehicle were Messrs. Pinchower and Fernbach, the gentlemen in borax, whose screeches exceeded those of the women down in the hold. At the first shot Pinchower thought he was a dead merchant, and straightening out on the roof of the stage, held to the railings with a very fair imitation of rigor mortis. As he made a good shield for the driver, that busy man did not disturb him. Pinchower at length sat up, examined his person, discovered no wounds, and promptly became very warlike. Fernbach, yelling "I'm looking for my gun," had dived into the front boot, where he continued to search the baggage until the stage reached Six Mile House. Thereafter, the sun making them unwell, the two men rode inside with the women and children, Pinchower occasionally feeling for a bullet he was quite sure he had received in the leg, or perhaps between his shoulder blades.

Ashim's trial at Independence lasted a week and drew on the flower of Panamint's citizenry for witnesses.

Dave Neagle and Jim Bruce, who had been in Darwin the day of the killing, pronounced it an efficient job though possibly a little heartless. Judge Murphy, the Panamint recorder, recalled that Perasich had been regarded as a difficult man whom Sour Dough Canyon had long been waiting for—an option forestalled by the event at Darwin. Pat Reddy developed to the jury's satisfaction that in the battle of March Eleventh everybody in the restaurant seemed to have taken shots but Montgomery; that Ashim's bullets all had missed, and had since been located in floor, walls and a customer's overcoat; and that, if this were denied, the dead man had been an unpopular fellow anyhow.

Under circumstances so understandably set forth, a typical Pat Reddy verdict seemed called for—and was rendered.

Ashim moved to other spheres of enterprise. He moved, it is true, at first stealthily and then with speed. For during the trial, Elias and Pete Perasich, who still seemed to doubt the quality of Inyo's courts, had again camped close to the north road out of Independence, waiting to assist his going.

Once more they closed on air. The stage rolled up and Ashim was not in it. But his one-armed lawyer was, ready for them with something more than an injunction and something for which the Dalmatian code had no immediate answer. Elias and Peter apologized and withdrew.

Bark Ashim watched the proceedings from the loft of Pat Reddy's stable. When darkness fell, he hitched up the faithful buggy and disappeared from Inyo County forever—presumably by the long road southward.

On for Havilah, Millerton and Markleeville rode Judge Reed, his hip-pockets bulging with hardware that was heavy and authoritative. Rumors that Pat Reddy had threatened his life could be set at naught—Reddy himself was suing the rumor-mongers, and the jury had seen to Ashim's permanent freedom anyhow. But a traveler never knew when he might meet a grizzly bear, and, besides, somebody *had* written him rather threatening letters.

Back at Independence, Reddy set to polishing his next brief.

## 23

## CHÁVEZ!

BELOW Kernville on the South Fork of the Kern River, a taverner named Scodie was taking his ease of a Sabbath evening. He sat on the porch before his little rustic store. It was February and the long Sierra valley leading up to Walker Pass was fragrant with sprouting oats and barley and pungent with wet sagebrush and the resin of pines.

Scodie had his shoes off. He twiggled his toes in satisfaction, for they felt free outside the leather and the news in his Havilah *Miner* was to his liking. Vásquez the arch-bandit of the California plains had been seized by sheriffs. Scodie remembered Vásquez. A dapper little man with bushy black eyebrows, who sang as he rode and ruled his men with velvet word and hand. A dangerous fellow—Scodie had been very brisk about setting out refreshment for him and his band when they once chanced that way. For the smiling, singing man was known for a ready cutthroat.

So now they had caught Vásquez, and he had been hanged.

At the account of how the wretch had wriggled on his San José gibbet, Scodie again twiggled his feet, and twiggled them once more as he read of the swashbuckling threats of Vásquez' lieutenant, written in bad Spanish and mailed to the judge and citizens. The lieutenant promised vengeance on "the just and the unjust alike." Cleovaro Chávez was the letter-writer. Tavener Scodie remembered this worthy likewise. Blocky, swarthy, heavy-handed—in spite of the fixed grin which somebody had carved on his face, a surly fellow. Too bad he roamed free. Well, Kern River in the Sierras was two hundred wide miles from Hollister in the Coast Range where the boaster had mailed his letters and whence, that day of Vásquez' hanging, he had mounted horse with great oaths and ridden away—

Scodie heard now the clatter of hoofs. Their approach had been muffled by the mud. Six or eight men were swinging from stirrup before his door. One had a slashing scar on his cheek that seemed to give his mouth a permanently evil smile. Scodie got behind his counter and asked what the travel-stained gentlemen might desire. With dismay he learned that what they desired was to cut his heart out.

In a flutter the host pushed his bottles forward and urged them to drink to Tiburcio Vásquez, that good, great, much-wronged man. It was, he perceived, a happy stroke. The bull-necked leader changed his mind about carving out the host's heart and smote him on the shoulder instead. The leader also ordered him tied up tightly, as was also done to a customer who came in; robbed the till of $800; appropriated Scodie's shoes; directed each follower to outfit himself with new hat, pants, shirt; left various old boots and evil-smelling duds in swap, took fresh mounts from the corral, and departed.

Scodie, with some genuine wriggling this time, finally worked from his bonds and freed his comrade. Together they sought out deputy sheriff E. B. Prater at Kernville. Ed Prater had a new Smith & Wesson revolver that shot rifle balls and could be loaded at full gallop, but in the chase that followed he had no chance to unlimber it. For Scodie fed his mounts good oats, and on those mounts the Vásquez men now rode.

A few days after Scodie's ill luck, the Panamint-bound stage passed Chávez' hard crew in the Cosos. There was a lively hustling of watches and wallets into hiding places under coach-seats and most of the personal weapons aboard were stealthily dropped from the off-windows. But the bandits were riding easily and waved them on, staying their distance.

Cruz López and José Guerra were natives of the Inyo sagelands who eked out an honest existence, when nothing dishonest offered, by mining around Old Coso and Cerro Gordo. Just now the pair were camped west of Borax Lake. Travel toward Panamint was picking up with the return of pleasant weather, and López and Guerra had it in mind to try a little private road-agentry. Hoofbeats coming up the road sent them to a crouch in the creosote bushes.

The horsemen appeared. One, two—Cruz López rose and leveled a cocked pistol. Three, four—Cruz saw with swift regret that he had miscalculated. Five, six—

The leader of the cavalcade sat in his saddle shaking with mirth. Such a burla: Cleovaro Chávez, successor to the immortal Vásquez, Chávez the gran ladrón, halted in the middle of the road by an impudent country bandido! As Chávez shook, López paled. There was a thrashing in the bushes and Chávez' comrades returned with the roped and stumbling Guerra, who had tried to bolt.

For a moment Chávez eyed the captured pair, then stopped laughing and bent closer to the man with the pistol. Yes, it was there, a wide triangular scar running down the fellow's right cheek from eye to beard. Everything was there—

thick brows, hewn grimace, blocky torso. It was as if he were looking in a mirror.

Chávez guffawed roundly this time, for here was material for both humor and strategy. What do you call yourself, hombre? Cruz López? Well, López, here is a scheme. You and your friend shall go free, and ride north while we ride south. Understand, López? Only, to everyone you shall meet you are Cleovaro Chávez. How they will wonder at the great Cleovaro striking in so many places at once! Before we part I shall give you some instruction in banditry that will make you rich. It is a bargain? Vámonos, then!

In the next few weeks, therefore, Chávez was seen and reported everywhere. It was an old trick of his dead chief's and had kept the western country in an uproar before.

Chávez was declared in Los Angeles, at the same time in Virginia City, at the same time far away on the coastside where he paid his mother and the Hollister grog-shops a visit and cried hot tears for the stretched neck of Tiburcio. Yet unquestionably, two days after he was seen by everyone at Hollister, he or one very much like him was drawing rein in Inyo county at Little Lake.

Leading four companions, the grand robber or his prototype entered that stage station, froze the keeper and his three helpers to stone with the utterance of his name, gagged and bound them expertly and cleaned the place out of horses. Next morning the northbound stage discovered the station men's plight and the rest of that vehicle's progress was a succession of meetings with robbed and foot-weary travelers who told feelingly of having met the brigands.

Word reached Lone Pine, fifty miles up the turnpike, that the desperadoes were surely coming. Jim Moffitt fell to worrying about a fine mule he owned. He put it up in Colburn's stable, which had tight doors, and made his own bed in the hay.

Toward midnight he was waked by someone fumbling at the latch. He blazed away through the woodwork and was pleased to hear a profane yell. He peered out through a crack and was still more pleased to see someone nursing a bullet-grazed arm. The wounded man was Colburn's partner, who felt certain that the shooter inside must be Chávez himself and no less. With his wound for proof, this partner ran for help. Half the town of Lone Pine and nearly all its shotguns responded, making an iron ring around the barn. Grimly they waited for daylight. When it came, Jim Moffitt, who was sure he had Chávez dead on the outside, raised his head cautiously to a window. Recognition was mutual.

During this siege of the Lone Pine barn another victim was being made of a sheepman in the Slate Range off in the opposite direction. This victim was relieved of $60, shoes and a horse. Then followed a descent on the stage station at Borax Lake and two other stations southward.

These visitations were probably conducted by the true Cleovaro, for there was an air about them. At the Lake he not only announced his celebrated identity but proved it by shooting the head off a chicken at a miraculous distance. In another instance he pleasantly offered to pay for meals, but was informed that the keepers made no charge to such courteous gentlemen.

Meanwhile the trails were astir with blasphemous travelers who came limping into the settlements lacking their money-belts, horses, and particularly their shoes. Stealing victims' shoes was done with a purpose, for a barefooted man turned loose in the desert was apt to be a long time arriving anywhere. That delayed pursuit. The true Cleovaro seemed to

be keeping south and east of Borax Lake, and the false Cleovaro north and west of it. The chief difference lay in the
flourish with which the real leader managed, and the small-
change avarice of his facsimile.

Among those to fall in with the false Cleovaro was one
Johnny the Frenchman, the place where he was halted was
sixteen miles down the grade from Darwin, the sum he
lost was fifty cents and both Johnny and his feet when they
arrived at the settlement were very, very sore.

Next was a man named Blankenship, who was deprived
of pistol and boots when he had twenty miles to go, and who
arrived at Darwin swearing that he had stepped on every
broken bottle and niggerhead cactus in the West.

Third sufferer, a nameless Indian, was done in more thoroughly, and his body tossed among the rocks and covered
with brush. Perforated by five bullets, he was found by
native trackers.

Fourth and fifth was the same man both times, Major
Thomas Hemming, an old Indian fighter who kept the toll
house on the Cerro Gordo road. Hut and person plundered,
the Major was tied up tightly and left in a most uncomfortable posture from which he finally twisted loose. Starting for town to rally its bravos, the Major had the bad luck
to run into his tormentors a second time. Angered at his
impudent escape, they now took him a distance off the road,
stuffed his mouth with his own kerchief, tossed him into a
barranca and left him there trussed so tight that the only
way he attracted the attention of the morning stage was by
raising one leg and holding it high—a most remarkable sight
in the desert. When the Major recovered his speech, he
swore that ten men had attacked him the first time and
twenty the second—five of the cowards having leaped upon

his back while he was holding the other fifteen at bay.

A few days later, W. L. Hunter of Darwin started on mule-back to visit his mines in Wild Rose Canyon. He told his friends he would be back in three days. At the end of that time he had not appeared. As his course lay over the same trail toward the Panamints where Johnny the Frenchman had been robbed, there were lugubrious regrets over Darwin's glassware for such a popular and worthy man. Later rumors that his body had been found inspired toasts to his memory. No confirmation of this tale arriving, it was concluded that perhaps, after all, Hunter lay tied up somewhere in the sage-brush, famishing and thirsting.

This sad possibility brought more rounds, resulting finally in the decision of four of his friends to go out and look for him. They rode forth stoutly, each robber-hunter armed with every weapon he could borrow.

Down-trail they came upon a solitary traveler who seemed to be having trouble with his donkey. The cargo of the beast was scattered over half an acre.

"Have you seen any of those damned greaser murderers? They're the fellers we fellers are after," pronounced the bold quartet.

"Yes, gentlemen, I *have* seen them. They just now went through me, and that's how this jackass's pack has got slung around this way. If you want to catch the black murderers, there they are, not a mile away."

"You don't mean it! Load up your jackass quick, and we'll stay and protect you!" exclaimed the bandit-seekers in one voice. And stay they did, until everything was neatly packed, when they valorously convoyed the beast and his master back to town. Hunter got in under his own power in due course.

To track down Chávez or—as men now perceived—to track down his double, the cavalry of Camp Independence was turned out. Captain MacGowan and a detachment of Company D rode hard for twenty-five days, scouring the Cosos and the country west of Panamint. Beyond Owens Lake they had a running skirmish with some fugitives and thought they wounded a man and a horse, but lost their quarry in a boulder-strewn ravine.

On the third day of May, William Dalle and Owen Coyle toiled into Panamint and paused gratefully at the Bark Saloon, first station at the lower end of town, for they too had a tale to tell.

Four nights earlier, while making camp with some teamsters at a waterhole seven miles north of Borax Lake, two Mexicans had materialized out of the gloaming with the drop on them and ordered them all to "heave up." One had a heavy scar on his face. Though not certain, because of the light of the campfire in their eyes, Dalle and Coyle thought they had seen six or eight other villains in the distance, sitting on their horses. With their three teamster comrades, the Panamint-bound pair had been tied up. The cook had been required to bake a big batch of bread. That, and all the other provisions in the outfit, together with money, watches, clothing, and of course shoes, the cook had been forced to stuff into sacks and fix upon a packhorse. The intruders had stuck to Spanish when they talked, and Dalle and Coyle were quite sure that if their own Spanish had been better they would have heard a very bloody doom for themselves being debated. It stood to reason the fellows must be murderous fiends who would turn mortal men loose in the desert at the beginning of summer with nothing on their feet but their stockings.

A day or so later the station keeper at Panamint Junction in the Argus Mountains was made guest of honor at a trussing-up surprise party. He was having these attentions all to himself when the stage from Panamint rolled up, whereupon its driver and passengers to the total number of six were similarly done by and laid in a row while the robbers sacked the establishment.

Bart McGee, one of the builders of the Surprise Canyon road, was next. He chanced to fall into the depredators' hands over beyond Shepherd's Canyon in the Argus range. Tightly corded, he lay low and listened hard. McGee had plenty to listen to and plenty of time to think about it, for he lay twenty-eight hours in a cramped position and heard his captors argue and re-argue whether to kill him quickly or toss him to a more lingering demise in a cholla cactus. Finally they told him they would let him go if he, on his part, would forget he had ever owned a horse. McGee retorted with spirit that he would follow them and kill them both if it took ten years.

The first night wore away. So, in secrecy, did Bart's fetters. But daylight revealed their frayed condition and they were renewed. Another long spell of constraint; then Bart, still pinioned, was put on a bareback mule and the mule dispatched into the open spaces with kicks.

Guiding with his heels, McGee got his mule to Darwin, where he claimed a new mount and set out, following tracks that led west.

Some days after, and miles away, Messrs. Spratt and Prewett chanced to look out the cabin window of their Chimney Meadows ranch, which lay on the Sierra uplands beyond Little Lake. Messrs. Spratt and Prewett looked out with interest, for two Mexicans were riding straight for their cabin with rifles advanced. It was a most impolite way to approach a ranchman's house.

Spratt drew his own Henry on the leader and leaned against the cabin door. When the rider declined to stop, Spratt fired. The visitors wheeled and fled. They were tracked down the mountainsides toward Indian Wells. Next day one badly wounded man was found on the edge of the desert. His rifle was raised and he was still full of business, but was rushed, captured, and hustled to a cottonwood tree. Standing in his necktie of rawhide he was induced to confess, and the confession showed that his section of the Chávez band had indeed never numbered more than two. A coroner's jury of Lone Pine citizens subsequently reported: "We . . . do find that the man was one of the robbers or bandits, and that his name was José Maria Guerra; that we found him in the canyon described, and that we buried him according to Hoyle."

On Guerra's person was found a memorandum book belonging to Bart McGee, which caused that stout-hearted desert man to wheel and gallop hard after the escaped companion. He came back in due course, whistling as he made his way down the canyon. He was once more riding his own horse.

Granite Station was a rude stopping point fifty miles below Panamint, operated by Messrs. Nichols & Littlefield. To this squalid shack, at ten o'clock on Sunday night the 28th of March, three men rode up, asked for barley for their horses and disclosed that they could do with a spot of supper. Mrs. Nichols bustled to this task. The travelers, who spoke broken English interlarded with Spanish, sat down and enjoyed their repast, then rose, drew pistols and with a certain courtesy invited the proprietors to be neatly packaged. They left with all the money in the place, the burly leader saluting Littlefield with "Adiós; you catch me, maybe," and were heard pounding north. Teamsters coming from that direction later untied the station folk and told of being passed by the trio, who kept three Spencer rifles and a shotgun trained on them while passing but made no move at molestation.

"They robbed 'a la chevalier'—no abuse, no violence, no rough overhauling in search of plunder," reported the *Guardian* when the details reached San Bernardino. "In the wilds of Inyo, splendidly mounted and armed to the teeth, their capture unless through treachery seems hopeless. Here is an inviting opportunity for our active sheriff to distinguish himself. We hope he will not allow the dashing Roland to monopolize all the honors."

But the sheriff of San Bernardino, with a county of size under his jurisdiction, sensibly pointed out that his northern boundaries were not monumented and that Chávez might properly be considered the guest of Inyo.

A citizen known as "Mr." Mowray ran an express wagon between San Bernardino and Panamint, carrying goods and passengers. Mr. Mowray, though occasionally given to exaggeration along certain harmless lines, especially if he had a tenderfoot on board, did not pretend to exaggerate the prowess of his horses. A day between watering stations, and a day or two of occasional lay-over, would bring him to Panamint, God willing, in a couple of weeks. So Mr. Mowray jogged comfortably along in early May with a single passenger. They left Black's Ranch, dragged up the heavy road over a rising wash into Black Canyon, gained the deep cool well under Pilot Knob so lately visited by Cleovaro and his mates, and set on to the twenty-three waterless miles toward Willow Spring.

The gateway to Panamint Valley between the Slate and Panamint ranges opened at last before them. As they entered that bald portal, Mr. Mowray's passenger, who was traveling to Panamint to negotiate for a store, a mine or a saloon, asked to be let down. Mr. Mowray, it must be feared, had been telling drivers' yarns.

"So they took this fellow's money and his boots," quavered the passenger, who had a good sum in gold about his person and a pair of excellent boots on his feet, "and made him walk to Panamint?"

"Yessir," affirmed Mr. Mowray, hitting a chuckawalla in its unblinking eye with a practiced spurt at a good dozen feet. "Twenty-five miles it is, from here to the bottom of Surprise Canyon, *and* six more to the top. Walked it in his socks, he did, right over the backs of tens of thousands of venomous square-shouldered high-chested tarantulas. 'Course they was dormant-like, it bein' the heat of mid-summer."

"I'll walk," decided the passenger, "while I've got my shoes on."

"Just as you say," nodded Mr. Mowray, who always was glad to relieve his horses. He bethought to add some more instruction concerning the fauna of the region. "You needn't be afraid of king snakes. They look awful but all they do is bite. No pizen. However, you might keep an eye peeled for side-winders, especially when they seem to be goin' on the ree-verse tack. They always come back a-fightin'. Then there's horned and gridironed rattlesnakes. I'll be waitin' for you at Willow Spring over the horizon yonder. If you sit down anywhere, *feel* the rocks, especially the gray ones. Tarantulas are gray too, but you can tell 'em from the rocks right away—they pinch easier."

Mr. Mowray jogged comfortably along, occasionally chirping to his horses and keeping his eye out for chucka-wallas. Does these tenderfeet good, he considered, to stretch their legs a mite. As he neared Willow Spring, with its odors of station-mistress Riley's cookery, the usually untemperamental horses shied violently. The blocky leader of three swarthy men had stepped out from behind a mesquite thicket.

It took the pursuing tenderfoot the better part of an hour to reach a curious bobbing object that had resolved itself on the desert ahead of him. The figure had been making progress by fits and starts, with increasingly more fits, it seemed to his follower, than starts. With caution the traveler drew closer. Yes, it was Mowray. He was stumbling along in his socks, bereft of wagon, horses, shoes and a pair of pants containing $128.25, and so generally unraveled that he missed a keel-backed lizard at four feet and responded to the tenderfoot's hail with only a snarling "Ya-a-ah!"

At Willow Spring they found chaos. Mr. and Mrs. Riley, a teamster named Ducrow, two foot travelers and an Indian buck lay bound in bale rope. Chávez and his gang, arriving on Friday, had tarried until the present Monday to assemble this human collection, though Mrs. Riley throughout had been "treated like a lady." Hadn't the leader with the hand-carved smile several times assured her, were his plans not so set, that he would positively have loaded her on a mule, packed her off into the sagebrush and dealt with her in a way most primitive? Mrs. Riley sighed. . . .

These doings down on the sandplains beyond its gateway did not leave Panamint emotionless. Robbery of its stage-coaches was not exactly a novelty. Travelers to and from that vale up there just under the skies expected to endure a few hardships. But all this marching and counter-marching, drawing steadily closer, seemed to have method.

Was it for reconnaissance? Did the swarthy rogues covet some especial jewel, some lightly guarded and particular treasure house?

Chávez and his late chieftain with their followers had repeatedly charged whole little California communities, sweeping through them like red tornadoes. The scroll was bloody: Suñol, where walls had been sprayed with bullets and useless murder done; Kingston where thirty-five men, tied and gagged, had watched their homes and places of business pillaged; Tres Pinos of heartless triple butchery; and only last spring, yonder on the near side of the Sierras, the sudden blast of rifle fire that had all but lifted the Coyote Holes station house from its log foundations. True, Tiburcio had been caught, entangled by a girl in the oak-shaded bottoms of Cahuenga Canyon, and subsequently lengthened in the neck. But the legend and the band of Tiburcio Vásquez still rode. And they rode close.

Imaginations up in Panamint readily concentrated on the idea that their town, perched above and behind nature's castellated walls and bastions, was the objective of this advance. Tiburcio's black-browed robber-knights, captained by the frosty-smiling Chávez, were moving to the escalade, the storm and the sack.

There was a flutter of preparation. A military plan was drawn up by Captain Messec, that stout old sheriff and Indian-fighter, and a lookout cabin was erected on a high outthrusting promontory. There was also a flutter, if guesses may be permitted, in the cotes of steeply ascending Maiden Lane.

Two residents of Panamint Mining District, coming down into town from their camp on the skyline, found the military arrangements in progress and eyed them with disfavor. "We'll handle this thing," said John Small and John McDonald.

Down in Panamint Valley Chávez continued, coming to camp at the hot springs a few miles beyond the entrance to Surprise Canyon. On the night of May 12th the pony rider with the mailbag from Lone Pine galloped into Panamint with the word:

"Chávez is coming!"

The pony rider had not seen the cutthroats, having crossed Panamint Valley just below them. But down the grade he had encountered a traveler at Sam Tait's station who had.

"He met them at the springs," said the courier, "and he hasn't got no shoes no more. Just blisters. He's swearing something awful."

"Come on," said John McDonald to John Small.

The two knights of the brass rail, properly belted, swung down the black-shadowed canyon. All Panamint that was still awake admiringly watched them go. It was fourth or fifth drink time when they set forth and lights were still making yellow squares of the tavern windows when midnight yielded to one o'clock, to two. . . .

Did Chávez really intend to test the gates of Panamint? Or did he look up at those looming walls and the long ramp winding steeply and darkly between; at the natural battlements that rose black and silver in moonlight, and would turn to something as hot as burnished brass by day? With a defending force lurking, perhaps, behind each natural tower, and a thousand towers to pass in that six-mile corridor, here was no helpless Kingston lying asprawl on a river-bank; no Suñol or Tres Pinos lying unsuspecting amid gentle hills.

Whatever Chávez may have intended in this vicinity, one look upward into those mile-high ramparts evidently sufficed.

Two and a half hours after midnight, Jack Lloyd's stage bumped into Panamint with the usual brake-squealing clatter. Driver and passengers were eagerly landed on for news. Had they seen the battle? Which way did its tide of fortune flow? Surely they had seen or heard something of those whip-tailed scourges of the desert?

Sure, J. McVey and Sol Ashim, the two descending passengers, had seen and heard plenty of the whip-tailed scourges. They were profane about it.

Chávez, then, had stopped them despite the valor of the defenders?

No, Chávez hadn't stopped them. They hadn't glimpsed hide or hair of Chávez. Chávez, if ever he had been there, was cleanly gone, and anyway was a gentleman. Small & McDonald had stopped them, affirmed the merchants. Had stopped them and ruthlessly taken over all their cash.

Fair enough, considered the Panaminters. Horatius at the bridge, or even two Horatii, ought to be allowed a little something for so much trouble.

# 24

# BILL STEWART LOOKS IN

THE Congress that adjourned March 3rd, 1875, found Stewart working with characteristic energy right up to the finish.

As chairman of the Committee on Railroads he denied to the Southern Pacific crowd a change of route to Cajón Pass, now definitely Jones' by capture.

Balancing the scales, he simultaneously helped the magnates thwart that other of Jones' pet schemes—legislation that would enable western states to tax the railroads' immense federal land grants.

Before Jones, who was chairman of the Committee on Post Offices, he then appeared and gravely argued for—and got—a post route from Lone Pine, California, to a place called Panamint.

The proprietor of Stewart Castle then turned his back on gas-lit Washington. If he left with regret, he also faced the next round of life's contest with abundant zest. Spring in the silverlands is a pleasant place, and Stewart was at the top of his powers.

He was on the Comstock early in April, gazing again into the voluptuous openings that had produced $243,000,000, and with especial awe into Con Virginia that now was tossing out the biggest monthly dividends yet divulged by any mine in American history.

Up and out of Nevada's holes continued welling the artesian flood of precious metal that was sixty per cent silver. Silver was the nightmare of the world. Gold was soaring, the price of its pale rival rushing downward. European capitals were in panic. Only in Virginia City and San Francisco was the universal dismay discounted. Everyone was rich, or about to be rich: the waiter, the barber, the housewife, the courtesan, the empire-builder.

As he looked on, stalwart silverite though he was, Bill Stewart may have had qualms. In spite of frequent market squeezes with their attendant transfers of wealth, their alternations of joy and despair, and their steady parade of suicides, the Pacific Coast cities were going increasingly speculation-mad. Fearful reckoning must be coming. Meanwhile it was a race between the Bonanza, searching for its metallic bottom, and the price of silver hurrying to drop beyond all bottom. Bill Stewart faced toward Panamint, uneasily reminded of a need for haste.

The eleventh of April saw him mounting the stage at Bakersfield. With him was George C. Gorham, California politico and secretary of the national upper chamber.

As the pair crossed the mountains they passed a stage bearing five travelers from Panamint. Stewart's coach pulled up abreast for conversation.

There was plenty of that commodity pouring out from the Bakersfield-bound vehicle. Brigands from the coastside were working a hundred-mile square of sagebrush for all it was worth and were striking everywhere. They had just tied up and robbed those men below Cerro Gordo, had sacked those stations on Borax Lake, and would undoubtedly be delighted to fall upon a couple of juicy statesmen and borrow their boots and wallets. Meanwhile did the travelers from the north have any extra cash and clothing with them, for these five travelers had been at Borax Lake when its stage station was raided—

The up-bound travelers advanced what they could, and pushed on. The station at the lake was a shambles. Fresh horses were scarce and the hostler was limping about with his feet wrapped in sacks. But the whirlwind had passed for the moment and the two visitors reached the top of Surprise Canyon with their own expensive socks still inside their expensive shoes.

Stewart found much to review since his departure of four months before. A million dollars, sown into that narrow high vale, had produced important changes. Tall chimneys and other structures had sprung into being. A large building of dressed stone was being made ready to house the company store. The *News* had doubled its page-size. Machinery for the big mill was arriving daily; its great boilers were on the ground.

On the night of his arrival the visitor was plucked from under the luminous yellow cone that shone in Dave Neagle's back room and induced to inspect that new iron-doored company storehouse. He was led away from Neagle's place mumbling protests—the draw might have filled a mighty promising hand. Still grumbling that he had seen plenty of bigger buildings with stronger doors where he had come from, he was led through the portals and a feminine and masculine chorus cried:

"Surprise!"

The interior was festooned with greenery and aglow with kerosene lamplight. A table of mill timbers had been rigged up to support Miss Delia Donoghue's cake and punch. Miss Donoghue and the other ladies of unassailable virtue from the upper end of town had been busy all day preparing for the happy moment. There were twenty-four misses and mesdames assembled for the function. Thus had Panamint expanded.

Into this gay awaiting mêlée the genuinely honored master of Stewart Castle waded gallantly. With Mrs. Storekeeper Chapman on his arm he led the promenade; with jaunty walk and lifting elbows he responded to Professor Martin's "All hands 'round" and "Forward to salute John Brown"; to strains of Old Zip Coon and Maid in the Pump Room, Haste to the Wedding and Chase the Squirrel the author of the Fifteenth Amendment advanced through buzz step, right-hand mill, balance-all, and gallop; while Panamint Tom, Indian George and Indian Jake, representing Panamint's first families, and sundry white men unequipped for social doings, pressed their faces to the windows in admiration. Long the strains lifted, while a spring moon rode up and out of Death Valley, swung low, kissed the roofs of the uninvited sisterhood down the vale, and sank afar into the Argus and Coso ranges.

For the next few days, Bill Stewart's presence again whipped everything into activity. Mails were lagging? The pony was set clacketing on his express route once more, this time with the pouches on the newly-won government line over from Lone Pine. Some of the boys getting kind of anxious to know when the big mill would start? We'll tie the whistle down hard, and there'll be jobs for all, lads, within sixty days! That concertina fellow who played for the ball the other night wants an interview on Panamint's future for the *News?* Well, then, tell him to paint it in any colors he fancies; he can't make it too bright.

Our New Era.—Panamint is about to enter on an era of prosperity [promptly recited the *News*] unequalled by any mining camp on the Pacific Coast. What with the large bodies of rich ore already developed; new discoveries daily being made; the excellent stamp mill of the Surprise Valley Company, now almost completed; that of the Sunrise Company soon to be erected; two others being contracted for; a railroad and telegraph under way, and soon to be completed to near our very doors, certainly ought to convince even an ordinary mind that we are not too sanguine or enthusiastic. But a few months will have elapsed ere the rattling of the numerous stamps in our canyon will produce an accompaniment to the reverberating sounds of the immense giant powder blasts now hourly heard. . . .

Accompaniment of another sort was provided sooner.

An evening stage pulled up before Neagle's place to take on its down-bound passengers. Stewart and his guest were preparing to mount to the interior. Just then Jim Bruce chanced to emerge from Dempsey & Boultinghouse's doorway and advance up Main Street.

Bruce's faro box had been behaving nicely, his coroner's business had been active of late and his claims were assaying better every week. In consequence of all of which, Mr. Bruce was whistling. The glad notes split the general nocturnal canopy.

It was at this moment that Bob McKenny quitted the interior of Bob Hatch's livery stable next above Harris & Rhine's store.

McKenny, a man in his strong young thirties, had been finding increasing difficulty of late in turning up the lady card and the oncoming faro dealer represented exactly the opposite of a cheerful sight to him. But there might have continued to be room for both if the winner in their recent encounters had only not been whistling. That shrill and unrestrained approval of things as they were amounted to just one more grievance than McKenny could stand.

In the light of the coach lamps his hand sped to his hip, and with serpent-like swiftness it came forth.

Bill Stewart had spent the better part of two-score years in developing a quick eye for just such situations as this. He dived now without dignity but with great speed from the far side of the coach, pulling the secretary of the Senate after him. McKenny's gun was beginning to spurt. The first shot tore through Jim Bruce's left wrist and knocked him to the ground. Two more whined through the night air lately made glad by Jim Bruce's song and the fourth embedded in the fallen man's back.

But even as he lay with the earth yanked out from under him, Bruce got his own weapon out and launched its contents at this man who disadmired his music. Two shots whined through the wagon spokes, behind which the passengers, one large and portly, one small and slim, were trying to make themselves of no size or thickness whatever. The next four all found effect, sieving Bob McKenny in hip, wrist, arm and leg.

Dr. Bicknell took charge of Bruce, operated on that durable man and pronounced that he would again be serviceable after rest and patching. McKenny was carried to Dr. Wells' surgery, where he died the next day. Editor Harris, who had come hurrying to investigate the fracas, arrived just as the distinguished travelers were emerging from behind the coach.

"Lively camp," said Gorham.

"Going to produce millions," averred Stewart.

"Self defense," adjudged Justice Smith, after hearing the available evidence.

Editor Harris waited until one or the other of the combatants was dead before selecting sides. He then dismissed the incident with:

"AN UNFORTUNATE AFFAIR.—We are pained to record that during an unfortunate affair which occurred at the express office, previous to the departure of the stage three days ago, one of our esteemed fellow citizens was compelled to resort to violent measures to protect his person. His opponent will be buried tomorrow in the little cemetery in Sour Dough."

No names, no cards, no flowers, no laments.

# 25

# THE ROSEATE HOUR

J. P. JONES did not immediately follow his colleague west. A new short session held him in Washington; in addition to which, he was having a superb time.

His gracious bride was the toast and belle of the capital. He had bought for her the Stanton mansion on fashionable K Street. His mines everywhere were pouring change into his waistcoat pocket. His popularity with his constituents was complete. He induced Congress to authorize a 20-cent coin in order to help use up the overwhelming output of western silver. He was about to be made chairman of a monetary commission to launch an exhaustive study of the uses and movements of the mysterious white metal. He was more than ever in a position to make things uncomfortable for the "big four" railroad barons. And his plans for reclaiming the southwestern deserts were moving.

Life was tasty.

Congress adjourned late in March. Jones still tarried, engrossed by large consummations. He invested in six blooded horses and ordered them off for the Coast in a special car attached to a fast train, shipping them at first-class passenger rates. Interested in that novel ice machine, and full of appreciation for the South's mint julips, he ordered artificial ice plants set up in Galveston, New Orleans, Mobile, Augusta and Atlanta. He bought rails, rolling stock and locomotives for his advancing railroad and ordered three palace cars—one for directors, two for passengers—all done in beefsteak red, the very latest word in sleeping car decoration. He commanded the presence of Chief Engineer Crawford, who came on rush summons from the Coast, and desired that work on the L.A. & I. be pressed at top speed.

In behalf of Santa Monica, now definitely taking shape in his mind's eye as the seat of his principality, he ordered a 2700-foot pier out to deep water; and to make sure that his pier got steamships, he set about making himself influential in the monopolistic Pacific Mail Steamship Company, of which his associate Trenor Park was a prominent director.

By the terms of this deal he also became a director in Park's investment in Central America, the Panama Railroad, with promptly stinging results.

Park and his isthmian road were angrily at outs with the steamship company over a division of profits from Pacific–Atlantic freights. The controlling stockholder of the Pacific Mail at that moment was considered to be the cold and sagacious Jay Gould. Gould also held mastery over the Union Pacific Railroad—the key to overland traffic. As Jones expected his L.A. & I. to connect with the Pacific Mail at his Santa Monica pierhead and with the Union Pacific across Nevada and Utah's vacant spaces, he undertook to cultivate Gould through a diplomatic maneuver. He would bring Gould and Park together and resolve their Central American differences. There was a meeting of all three at which Gould not only agreed to yield on the Pacific Mail matter with Park, but bought heavily into the latter's isthmian railroad and advised Jones to do the same.

Pleased with his success as a handler of financiers, pleased

to have Gould for an ally at both ends of his L.A. & I.,
pleased with the blow his L.A. & I. was now in position to
deal to Huntington, Stanford, Crocker and Hopkins—
pleased with everything—Jones consented to take a little
flier in those isthmian shares. He had the happiness of seeing
them vault a dozen points in a day.

"Didn't I tell you? Better buy more," exhorted Gould.
Jones bought more.

Whereupon, complications. Jay Gould, it developed, when
he glibly committed the Pacific Mail to Trenor Park's de-
sires, did not possess Pacific Mail. Jones' arch-enemies, the
Southern Pacific crowd, all unknown to everyone but Gould,
had seized control of it. Now, crying that Park had shown
favoritism to English ships, they repudiated Gould's treaty
and threatened Park with suits for damages. Down rushed
the Panama rails which Gould had been feeding out to
Jones as fast as Jones wanted them. Before the Dustman
could let go, his fingers had been nipped a painful $700,000
worth. Simultaneously, the Southern Pacific brought its first
train tooting for freight to the base of the Tehachapi Mountains, and that fine new teaming road up into Inyo County
began to bring it all upland traffic.

While Jones had been concerning himself with somebody
else's cargoes across Panama, the enemy railroad barons had
neatly assumed control of the commerce of his own deserts.

Jones turned west in June. His tilt with Gould had pinked
him sufficiently to force fast disposal of his St. James Hotel.
But his mine at Kernville was remitting $165,000 a month,
other prospects were flourishing, and Bill Stewart was send-
ing him assuring news from Panamint. Arrived at San Francisco, the man of many interests was soon pronouncing:

"In the development of all noted mines on the Pacific

Gabriel Moulin photo.

"The huge monster started gracefully."

"Took to sitting on the grassy slope and watching its black smoke."

"The rocky walls all but touched."

Coast, very rich surface rock has been found and then a barren zone was encountered. This had been the case in the celebrated California mines and in the Comstock, and it has also been that way in Panamint. We have gone down to a depth of three hundred feet and have found rich and reliable veins. The outlook for Panamint cannot be better."

While workmen rushed in the bright spring weather to complete the Panamint mill and roasting furnaces, the townsite at the other foot of Senator Jones' rainbow was likewise preparing for its great moment.

Discarded long since was the name Shoo Fly. After tentatively trying on the name Truxton, the future seaside metropolis had decided that its proper appellation was Santa Monica. Far into the ocean now, nearly half a mile out, jutted the trestle which was to gather in the steamships and dispatch Panamint's ores to the world's far smelters. Down the Atlantic, around the Horn and straight for this wharf was driving the ship "Kalorama," her hold freighted deep with iron for the Senator's railroad. In June the newspapers of San Francisco exploded with announcements of a great sale of lots proposed for mid-July on the heights behind the trestle.

These advertisements fell upon Los Angeles with dismaying force. "The arrival of San Francisco papers of the twelfth instant, containing an advertisement announcing the sale of property in Santa Monica, and asserting that that place is destined to become the great commercial city of Southern California, has produced a profound sensation here. Residents regard this as an effort to destroy Los Angeles," cried dispatches from the southern metropolis.

In spite of which neighborly qualms, the day of the great land sale was awaited with excitement.

It broke with a gorgeous dawning. Those citizens of the southland who drove their buggies through the dust from Los Angeles, scattering bands of sheep as they moved, were greeted with a magic view as they crossed the low divide.

Before them the blue ocean lay clasped in horizon-reaching arms. Little rollers, come to the land auction clear from Cathay, broke musically on the shore. On the north the Santa Monica Mountains, sweeping down from the high San Gabriels, waded out to meet the breakers. To the south Point San Vicente formed the other horn of the bay. A single great oak at a fork of the road directed traffic toward the auctioneer's stand, which was perched on the bluff. Several hundred people gathered early around it. The steamer from San Francisco came in trailing its smoke, docked at the long pier, and discharged more multitudes just in time.

The new townsite was planned on a generous scale. Its streets were 80 to 100 feet wide, its Ocean Avenue was laid out with a width of 200 feet, and space was marked off for a future plaza, hotels, a young ladies' seminary—even a university. Though the only visible improvements of the moment were a single board shack, the wharf, the beginnings of a frame hotel and a scattering of tents, here was grandeur in the making, and an orator mounted the stand and made the fact plain to all. Declaimed the Honorable Tom Fitch, late of Washoe's bar, over the top of his lemonade pitcher to the picnicking hundreds:

> Today on Wednesday afternoon at one o'clock, we will sell at public outcry to the highest bidder, the Pacific Ocean, draped with a western sky of scarlet and gold; we will sell a bay filled with white-winged ships; we will sell a southern horizon, rimmed with a choice

collection of purple mountains, carved in castles and turrets and domes; we will sell a frostless, bracing, warm,
yet unlanguid air, braided in and in with sunshine and
odored with the breath of flowers.

The purchaser of this job lot of climate and scenery
will be presented with a deed to a piece of land 50 by 150
feet. The title to the land will be guaranteed by the
present owner. The title to the ocean and the sunset,
the hills and the clouds, the breath of life-giving ozone
and the song of the birds, is guaranteed by the beneficent
God who bestowed them, in all their beauty and afflu-
ence, and attached to them in almighty warrant is an
incorruptible hereditament to run with the land forever.

With which rousing send-off the town was launched. Two
months later it had a hundred houses. On October 17th,
1875, the first train over the L.A. & I. thundered from the
pierhead as far as Los Angeles. Not yet arrived were the
elegant coaches, and passengers were accommodated on flat
cars, but three trips were made that day in what everybody
agreed was considerable dash and go. By November, Panamint's seaside counterpart had a newspaper, telegraph office,
two hotels and a school.

"Well, I think I am able to say that I have secured outside capital to cover my 'blind,' to use a phrase familiar to
betting men," Jones said to an interviewer. "The line between Los Angeles and Santa Monica has been pushed forward, and satisfactory advances have been made on the
tunnel at the Cajón Pass. I pay as I go. I do not pretend to
be able to build a railway out of my private purse, at once.
That would require too heavy a tax for a man who has so
many irons in the fire as I have. I could very readily
borrow the money to complete the road at once, but that

is not my way of doing business. I have no doubt but that Stanford and his friends would be glad to advance me the means themselves, in the hope that I might be led into deep water, and by calling for everything at once, they might get the road. I commit no mistakes of that kind."

It was a reassuring statement and timely, for of late there had been that less assertive tone in conversation at Panamint, and on June 14th the usual optimistic *Alta* at San Francisco had reported: "Letters from Panamint are less sanguine than those written three or four months ago. . . . There is much rich ore, but the quantity of the richest quality is less than was anticipated, and the refractory character of the ores, and the lack of facilities for reducing them on the ground or of shipping to Europe, stands very much in their way. The expectation that there would be a rush of miners to the district last Spring was not realized and there is at present much doubt whether there will be any considerable growth until after the proposed railroad from Los Angeles to Panamint has advanced twenty miles or more beyond Cajón Pass."

But with Jones serene and confident, dirt flying in the Cajón, iron rails actually arriving, and the Big Mill completed and ready for firing, Panamint had made July 4th, 1875, a day of days.

Meat-dealer King had been again called upon for his cart, this time for esthetic purposes. A certain Grand Marshal Paris, assisted by a civic-minded Mr. Stebbins, decorated it as a car of splendor and installed within it a white-robed young lady as the Goddess of Liberty with a supporting cast of three little girls. Drawn by human steeds and preceded by a band—one drum, one tuba—the float made a triumphal tour down Main Street and was lifted by hand and turned around, goddesses and all, where the gulch grew narrow.

Smoke blacker and blacker was pouring from the Surprise Company's tall new stacks. With the return of the procession, Superintendent Messec pulled the lever of the whistle down. While the mountainy walls of slate, limestone and granite caught and hurled back the blast and the juniper-fired boilers quivered, the engineer slowly applied the steam. As on a test run five days earlier, the huge monster "started gracefully." To screeching whistle and rattle of ore cars on the overhead tram, thirty stamps began their dinful and eccentric dance.

At this historic instant a banner was run up on the flagpole fronting the Surprise Company's store. Miss Delia Donoghue, with her cheeks bright from cookery, watched this ceremony from her porch next door. Cheers greeted the gonfalon—cheers that changed to a delighted yell as the article shook itself out into the breeze.

It was no starry bunting, this battle-torn object, though it swung aloft with a certain commanding presence.

The imposing apparatus was open at the top like a majestic urn. From this sumptuous mouth the Vase of Venus swelled downward at the front into notable double bays; struck inward at the sides to a vise-like middle; flared curvily out and descended to a broadly oval base; then dipped at the central front to an implacable point. Though empty now of fair flesh, the magnificent container was full of reminiscence. Shouts from a couple hundred throats greeted it.

The up-canyon wind blew the female torture-chamber slowly around. That evoked more happy recognition. For now there appeared behind the contrivance a big bolster-like attachment of wire, whalebone, crinoline and assorted stuffing, hanging with a bulldog grip. This curious afterwork was surmounted by another of horsehair and muslin, further revealing how Dame Fashion achieved her prevailing and astonishing dernier.

Miss Donoghue, who had stood rooted to her spot, at that revelation picked up her skirts, fled inward to her Wyoming Restaurant and slammed its door. *Those men!*

It was just as well that the camp's chief priestess of respectability stood beyond view of the throng, pressing her hands to burning cheeks, feeling around at the back for her own betrayed mechanisms, and doing her peeping from behind a curtained window. The girdle of Hippolyta kicked and danced at the top of the halyards to continuing cheers and the full-voiced steam whistle; danced wantonly, sprang open at the front, and came to rest only when clapped wearily around the flagpole.

It was the sign affixed on high to the hellish woman-trap that completed the undoing of Miss Donoghue. The placard, which stayed there until dusk, proclaimed:

"This is the flag we fight under!"

That night there was a big keno pool at Dave Neagle's; and a calico party up the grade—each lady depositing a hand-made necktie at the door and wearing a dress to match it, by which her swain identified her. Miss Donoghue had gone to a deal of trouble over this occasion. Farther down the hill, around Neagle's corner and up the steep and narrow way of Maiden Lane, the ninety-ninth anniversary of the Republic was also celebrated—in this case with a garter party, affording unlimited practice in the doctrine of freedom to those who sought their partners.

At Miss Donoghue's fete the frolicking ended shortly after midnight, but at Martha's, where that lady was once more restored to her corsets, the rejoicing didn't subside until dawn.

Two residents of Panamint were late for these assorted exercises; in fact they did not arrive at all. July weather might be balmy, even exhilarating up there in the high-swung dale, but it was the breath of death out on Panamint Valley's floor. Albert Quigley, a youthful miner, and Oscar Muller, proprietor of Panamint's Arcade Saloon, had been visiting in the Cosos. They set out independently to rejoin Panamint in its day of carnival and they both misjudged their strength and the capacity of their canteens.

Days later their bodies were found at points far separated. Both were face down on the sands and both had died with fingers clawing frantically for water.

Meanwhile Charles Crocker of the Southern Pacific was writing to his associate, Collis P. Huntington:

"I notice what you say of Jones, Park, etc. I do not think they will hurt us much; at least, I would rather be in our places than theirs. I will ventilate their 'safe harbor.' "

To which Huntington was bleakly replying:

"I shall do my best to cave him [Jones] down the bank."

# 26

## THE COMING OF THE WATERS

PRODUCTION at the Big Mill began in earnest seven weeks later. The ore that rumbled down from the Wyoming and Hemlock tunnels was smashed to a pulp running $95 to the ton—no Consolidated Virginia, but quite a respectable showing. More than one pair of eyes watched the stiff-legged dance of the machinery with satisfaction. Soon the rich essence of the Panamint hills would be moving down-canyon in virtually cash form, and definite plans were being laid around its going.

Wells, Fargo & Company still declined to have anything to do with Panamint's knobby canyon and knobbier residents. This decision by the express concern that owned more guns and ammunition than a Balkan army, and had carried on running war with the West's whole profession of road-agentry for twenty years, had done much to establish the camp's reputation as a place apart. But it did not solve the immediate problem. The more Bill Stewart thought about that issue the thornier it got.

John Small and John McDonald were not in a hurry. They had Wells Fargo's absolution for the past tucked snugly in their pockets; they had plenty of stimulants, plenty of tobacco and plenty of time. Along with their mineral claim they had passed over to Stewart & Jones all mundane worry, and now took to sitting on the grassy slope opposite the mill and watching its black smoke. The thunder of its stamps was music.

"Seems like it's working for us—just us," said John Small.

"It is," affirmed John McDonald piously.

The 2600-foot cableway sent down its buckets in endless parade from the mouth of the Wyoming above, and from the Hemlock which connected with the upper end of the tram by a mule road. At each descending load of high-grade rock, McDonald and Small felt entitled to grin more broadly than ever.

"When are you shipping the silver out?" the two Johns asked Bill Stewart once or twice.

"Pretty soon, boys. I suppose you'll miss it considerably —seeing how long you've been sitting over there and watching."

Jested John McDonald: "We won't miss any more of it, Senator, than we can help."

Acutely aware of that single wagonway, six thousand feet deep and tortuous as a blacksnake, Bill Stewart nevertheless joined in the general laugh.

The tramway went on rattling and discharging. The tall stacks belched. The boilers shook. The thirty hammers merrily rose and fell. The furnaces smoked and roared. The concentrating vats gurgled and swished. The loiterers lounged and waited. With early autumn:

"Let's be going down the hill," said John McDonald to John Small.

About four miles below the town, though still a couple thousand feet above the desert, the rocky walls of the canyon all but touched. Here a man sitting on a wagon could stretch out his arms and nearly brush both sides with his fingers.

For twenty and thirty feet above his head, the cliffs were worn by the waters of old storms. Higher, the walls widened out to let in a wash of sunlight; but here all was shadowy even at noonday.

It was not noonday, however, but stygian darkness in the slender passage when two men put out their fire, extinguished their pipes, and dropped behind a bulwark of stones.

The living, teeming alertness of the desert night surrounded them. The infinitely stealthy pad of the kit fox searching for his supper. The alarm-note of some kangaroo rat at the bottom of his burrow, whirring his strong hind feet against the sand. The sustained cricket-note of a grasshopper-mouse in his hole beneath the rocks. The piping whistle of the long-eared, lumpy-nosed burro bat as he darted on swift wing. A sun-loving tarantula, big as a saucer, found himself deprived of his hole and withdrew ball-like into his fur in deadly fear of the night-chill. His private doorway, equipped with an ingenious trap-door, led to a warm gossamer-lined nest which he much desired; but the doorway was barred just now by a ponderous object.

So passed the moments, vibrant for all their stillness, shared by a myriad of living and unseen things. The spiny pocket-mouse, no bigger than a pipe-bowl, hopping a couple of feet at a jump and stuffing his cheeks with seeds. His tiny cry at the lunge of an on-creeping side-winder. The deep, rhythmic chuckle of a watching horned owl. A millipede, following an all-day battle with a scorpion, rolling over on his back, playing dead and suddenly lashing out with the victorious coup de grâce. The death-battle of a grasshopper to save its young against the mate of the same armored enemy: dodging, keeping between its babies and the tank-like giant, lunging swiftly at last and biting off the stinger of its foe. . . .

"Seems like," said John Small, bored by the absolute nothingness of his surroundings, "that wagon ought to be comin' along pretty soon."

The outgoing silver, reason had told them, would be shipped in darkness. Reason also told them that it would not move lightly guarded. Well, it would not be attacked exactly lightly, either. Small and McDonald were shy on immediate provisions, but not on armament.

But night rolled away; fox, bat and owl returned to their holes; murky dawn sought and found the canyon depths and its stiffened tarantula— "Hell," said John Small as he flung away the distasteful chip, "I bin sittin' on it"—and the flasks of the pair had long since been drained; but still no wagon.

The hour for breakfast came and went. So too the hour for noonday dinner. There had passed only conventional travel. Small and McDonald were hungry, sore in spirit and excessively cramped.

"Seems like," said McDonald, "the Senator has disapp'inted us."

Day at its two-o'clock brightest, and the canyon-bed actually gilded by sunlight, brought at last the clamor of bells to their ears.

"It's Nadeau's freight," said Small and McDonald, and they crouched behind their rock.

"Howdy, boys," said the driver when they stepped out. A nameless hero, he had been picked for his cheerful willingness to face an almost certain pair of shotgun muzzles. He was facing them now. And the muzzles looked as big as the bottoms of a pair of overalls. All but two of his mules were hooked on behind. Beside him on the high wagon-seat there was neither companion nor weapon.

"Where's that silver?"

"Right aboard. Senator says to tell you boys to do what you think best. Says I got enough to do tryin' to outguess mules, and not to start matchin' wits with you fellers."

Small held the gun and McDonald stalked back and climbed up over the tail-gate. There it lay, a gleaming ton and an eighth of virgin metal. It was present, he saw with astonishment, in just five big balls. Tentatively John McDonald tried rolling one, while the driver hid his smiles in his hay-like whiskers.

Five 450-pound cannon balls of shining silver are substantial property in anybody's town. But they are not worth a pint flask or a ham sandwich in the bottom of a steep-walled ravine, three-score miles from the settlements, with nothing intervening save a sea of deserts. Even ousting the skinner and driving away with the wagon gave no solution. Faster than wagon could jounce would travel the word of its coming —and no two robbers could fight the world, even with 450-pound cannonballs. Or spend them. Or get change for them.

"Senator says," transmitted the driver finally, "to ask, hev you et? All right, take my victuals. Sorry this consignment ain't in the shape you expected. Guess I'll be getting along."

Small and McDonald stalked glumly back to town and told Bill Stewart, in the forthright and earnest language of their forefathers, exactly what they thought about him. But, as the Senator pointed out, they now knew merely how the rest of mankind was feeling, that October of 1875.

For the whole world, as well as Small & McDonald, was bothered by too much silver. It was a repletion of which there seemed to be no end.

THREE HUNDRED miles to the north the Comstock kings were driving their workmen in all haste to clean out the

Bonanza before Sutro and his tunnel should win access,—before mine timbers should crumble in the underground damp and heat,—before nations should renounce the metal that was making universal wreckage.

Around every stock board, hysteria continued. Throughout spring and summer this fever had transcended anything previously known. The ups and downs of the Comstock had the cities of the West atoss like hotcakes on a skillet.

A shrewdly engineered pinch in April, based on a hint that the Bonanza was playing out, squeezed from the values of the Comstock mines twenty millions in a week. This cracked the important brokerage house of Duncan Sherman & Co., which among other and larger affairs had been trying to sell for millions a few hundred acres of the silt of Panamint Valley. July brought another disturbance, a sudden drop in Idaho mining shares, and occasioned such a tumult in California Street that extra police had to be called out to clear the sidewalks.

These frequent upsets to popular feeling were further caused by a Comstock phenomenon: the steady rush of hot water into the emptying vaults.

Whence came those steamy rivulets?

Mysteriously, relentlessly, they trickled and sprayed in. They pressed their attack from the east on the 2,000-foot level. Night and day the big pumps labored with this threatening deluge out of Avernus. Men in underground workings were forced to work in short shifts; they were brought up swooning from the labyrinthine bathhouse. At deeper levels there was perpetual struggle with cold waters. Men were used to that. But not to these hot seepings.

Then came August, and more trouble.

It started, as all recent crises, with the rumor that the

Bonanza was exhausted. Were the Bank Ring and the Comstock's new rulers once more trying to frighten each other out of their holdings? Were they leagued to plunder again the ten thousand small fry? Or had the fissure, which nature had spent thirty million years in filling, really been cleaned out in thirty months? Fable spread dismay. Dismay spread terror. Con Virginia once more plunged. For every $10 it had paid in recent dividends, now it threw off $50 in value. Companion shares tottered. Trapped by the fascinating game of margins, people rushed to unload. Such scenes had been enacted before. But not with attendant earthquake motions in the base-block of the Coast's financial citadel.

On Black Friday, August 27th, three days after the quartz mill in the distant Panamints settled down to business, the officials of the Bank of California called their manager to an accounting. Then it was that Ralston—Ralston the Magnificent, the entertainer of princes, the refashioner of cities, the sponsor and backer of Sharon—the man who had once by a snap of his fingers called up a million in milled money from the United States Mint without security—was discovered to be short in his accounts and disastrously over-extended. The mightiest fiscal institution in the West closed its doors with a clang. A few hours later Ralston was fished from the Bay, a corpse.

Again it was the familiar story. "Opulent individuals but a few months ago are now penniless debtors. The homestead is turned over to satiate or appease the rapacious demands of creditors, or has been sacrificed to the inexorable calls for margins. The demoralization embraces all classes of men and all branches of business." There were several pistol shots. Among them was one fired in Los Angeles, where the smash-up of the Bank of California forced Temple & Workman's bank to close. The suicide of his partner was a finishing blow to F. P. F. Temple, the father of Senator Jones' Los Angeles & Independence Railroad.

Forgotten as any other chip in this general whirlpool was the camp in the hanging valley of the Panamints. But it was sucked and heaved and tossed around with all else.

Jones insisted he was not affected by the suspension of the Bank of California. He announced that he did not owe the bank a dollar, nor did it owe him. Stewart also assured friends that he was bloody but unbowed. No longer in Congress, but still the Voice of Silver, in a ringing public letter he charged the ills of the hour to a shortage of money. Speculation and overextension, among high and low, were not yet seen to be the basis of it all. It was still Overproduction of Silver, in the opinion of the gold men; it was Demonetization, the "Crime of '73," in the minds of the silver advocates.

"Let the mints be set to work," cried Stewart, "coining standard silver dollars, and when coined use them to pay debt according to contract. This alone will reduce the price of gold until the gold and silver dollars shall again be equal. This would restore property to its real value, secure the development of the silver mines and save the country bankruptcy."

But that ancient enigma, the ways of money, which would continue to defy solution for decades to come, was not going to be resolved in time to rescue a nation from hard times, or a thirty-stamp mill and a collection of huts in a mountain gulch from irresistibly approaching extinction.

For winds of several kinds were now rushing to put out the little light set burning in the Panamints three years before by Jacobs, R. L. Stewart and Kennedy. World-wide depression, that was starting even as the camp was being founded, might have been survived. Political assaults on silver, over-production of that metal and the great fall in its price, under happier local conditions might have been ignored. Even the Bank Panic of August, '75, shaking $42,000,000 out of the market worth of Virginia City mining stocks and under-mining the values of everything else, might have been out-ridden. But there were other hostile forces.

Prime among these was the uncompromising toughness of that Surprise Valley rock. The second was a slowly dawning conviction that Panamint's gaudy cliffs after all might hold but surface glitter.

So by this time Panamint knew. With the wisdom of all mineral camps, it knew.

In September Panamint was hearing that its all-important railroad, with a tunnel already 300 feet into the Cajón, had temporarily stopped work for lack of funds. A month later, the Southern Pacific slashed rates on its spur between Los Angeles and Wilmington. That was a financial staggerer to the parallel Jones line. The caving of the Silver Dustman down the bank was realistically under way. Temple was smashed, his partner a suicide, everybody broke and stock subscriptions to the L.A. & I. impossible of collection. There was nothing for the L.A. & I., that lifeline to Panamint, to do but wait for the clutch of slowly reaching Southern Pacific fingers.

And Panamint's two angels, despite their denials, were too shorn about the wings and plucked in the pin-feathers to do anything about it.

Then, to make desolation out of disaster, Virginia City toward October's end caught the sleazy fabric of her gown in somebody's kerosene lamp and the whole town flamed high and went black in one vivid, garish, windwhipped night.

Fire above, water below. The destruction at the surface could be remedied, but not those deeply enroaching trickles, sprays, floods.

Inscrutably they continued playing into the candle-lit gloom. Silently, mysteriously they came. And when they came, men left. The pumps were working overtime now. Should they falter, the world would no longer be menaced by a deluge of silver.

The great mines went on producing as never before, despite fears for their future. The stock panic was merely one more quiver and readjustment in a topheavy speculative structure.

But meanwhile the rising hot waters mingled with icy flows from other levels. Dry old Mount Davidson was bleeding inwardly at every vein.

AMONG those dwellers at Panamint who never argued with the inevitable was Billy Killingly. He mounted a well-chosen horse and led forth a string of packmules to whose selection, if not to whose ownership, he had given professional attention.

Up toward the crown of Telescope rode Billy, halting for the night at the top of a ravine on the bold west face. In the folds below twinkled several small fires whose gleams he chose to avoid. In a high pasture browsed a pair of horses which Billy felt would look well in his string. Day broke early, but not early enough to witness his departure, which meant rounding the big mountain, striking down Blackwater trail for Death Valley and making for the Funeral Mountains opposite.

Out across the superheated gravel and salt and up through Furnace Creek Wash pushed Billy Killingly. Dawns found him making camps at discreet distances from the water-holes;

dusks and the stars found him swinging to saddle again. Travel by night is the way to escape the desert sun. But it is not the way for observing matters behind one's shoulder. In consequence the Pioche *Record* presently pronounced: "BILLY KILLINGLY KILLED.—A horse thief and desperado of that name, who lately figured about Panamint, has been killed by two men from whom he stole horses. These men followed Killingly some 300 miles, recovered their horses and left the thief's bones on a desert not far from Hiko."

Among those who moved on and out of Surprise Valley was Isidore Daunet, accompanied by five other disappointed rainbow-chasers. These men quitted Panamint by the divide at the back. Beyond the malignant pan below, on a straight route if they dared to take it, were the Desert of the Colorado and the rumored treasure fields of Arizona. The seven took the dare. They plunged into sizzling heat and three perished. Daunet and two companions barely pulled out alive. But they pulled out with more than life itself. They returned to camp with samples of "cotton ball" borax, and the result of their discovery was the Eagle Borax Works, first enterprise of that sort in Death Valley.

Among those who moved down the windings of Surprise Canyon was James Bruce—painfully, for his going was accomplished in a spring wagon made into an ambulance. Bruce was tough, but his system had not thrown off Bob McKenny's lead balls and he was being shipped to San Francisco for an operation.

The fall election momentarily turned back the clock. Though many of its characters had left for Darwin, Panamint still had 302 names on the great register to Darwin's 57, and about a hundred of the prodigals came home in November by stagecoach, wagon and deck of mule to vote. The cause of it all was a bold attempt by Dave Neagle and Rufe Arick to cut off and set up a new county, "Panamint," from Darwin southward and rejuvenate Panamint town by making it a county seat.

That night the lamps once more burned brightly in Neagle's Oriental, Joe Harris' Occidental, and Fred Yager's Dexter, though Oscar Muller's Arcade was dark with the darkness of perpetuity and many another once-jolly fire went unkindled. Up Maiden Lane there was notable business. But on the following morning, when it was ascertained that Postmaster Swasey had replaced Will Smith as justice and that Inyo was not to be cut into two counties, exodus resumed.

Down the canyon this time, queenly as on the day she had come, passed Martha Camp. An imposing assemblage of furbelows, whalebone, horsehair and high heels, she ambled down the flaggings of the wicked street and with helping boosts gained the box of Jack Lloyd's outgoing stage. Her handmaidens were elevated to the deck behind her, and there they were, Darwin-bound and gay as ever, their chignons, shepherdess bonnets and hard-bright little souls only a trifle the worse for wear. Though when he saw them leaving, Joe Harris for one couldn't bear the thought of that empty Lane. He too swung aboard, homeward bound for Independence. The door of his Occidental was drawn closed but he did not bother to turn the key.

There were occasional flashes from the guttering candle. In February of '76 an eight-foot vein was struck, 400 feet deep in the Hemlock, that looked better than anything previously discovered in the district. "The company," came a characteristic report, "does not expect to stop short of 1,000 feet, or another Comstock." The big mill found itself crunching $50,000 worth of ore a month and netting $20,000. Jacobs' little mill, leased to Dave Neagle, was likewise thumping. To those of true faith, the bonanza that would yet startle the world was still just a turn or two distant, a spadeful or two below.

So a number of the Panaminters stayed on, though mail had been cut to once a week and stages had curtailed schedules. Harris & Rhine continued its mercantile establishment, with tall, thin J. A. Brown in charge. Dave Neagle continued as the camp's chief cheer-dispenser. Captain Messec went on exploring, delving, probing for that second Comstock that could not be far away. But in that spring of '76, Dave Neagle sniffed blue-blooming sage and yellow cassia on the breeze that was coming up the canyon, and could stay no longer. He packed a mule or two, saddled a riding horse, turned his keys over to Jim Brown in case Harris & Rhine felt like enlarging, and took the long trail out—a trail that led to many places, including Bodie and the uproar that was Tombstone and the civil commotion that was Butte, Montana. The trail never led back to Panamint.

Small and McDonald dropped into town soon after. They found things depressingly quiet and decided to stir them up a bit.

At half past seven on the morning of April 20th, Jim Brown handed over the affairs of the store to his assistant, young Faulkenthal, and went to his breakfast. He left Faulkenthal and ex-Justice Smith warming themselves beside the stove.

Life had not been serene of late for ex-Justice Smith. Homicides had ceased, but he was lamenting to the other stove-warmer about the recent theft of a pair of his own mules, a matter which he felt Small and McDonald could explain. Small and McDonald had been hearing something of these aspersions and were sitting across the street in Coulter's place, keeping armed watch on Harris & Rhine's through Coulter's screen door. They saw Jim Brown go out, accompanied by the justice's young brother.

Faulkenthal and Judge Smith continued to toast and to talk. A customer came in. He bent down, appearing to be looking at a case of boots. Clerk Faulkenthal got up to serve him. The customer straightened and it was John Small, with a revolver in his hand. The square-built man gestured clerk and justice to a prone position on the floor, and in a jiffy the two men were trussed back to back in a way that would have delighted Cleovaro Chávez.

Jim Brown and the justice's brother, comfortably break-fasted, came in just then. John McDonald confronted them in full fighting regalia and added them to the bale-rope party. The same went for a customer named Pete and an in-quiring Chinese. All were tossed into a sleeping room, the case of boots was piled atop them, McDonald stood guard with a magazine rifle, Small emptied the safe of $2300, and the rogues then left. And this time they were leaving for good.

The cliffs of the precipitous gorge ran together behind them. The desert plunged away at their feet. Will Irwin, a miner riding over from Wild Rose Canyon, met and saluted his old acquaintances as they passed. They moved jauntily, as befitted men who had plenty of cash and all the world stretching out before.

Long before, Editor T. S. Harris had harked to the call of the horizon. One by one his advertisers and subscribers had withdrawn. Many of them, like Ned Reddy and Perasich, had taken their affairs to Darwin. To Darwin then, queen of the Cosos, Harris concluded to remove.

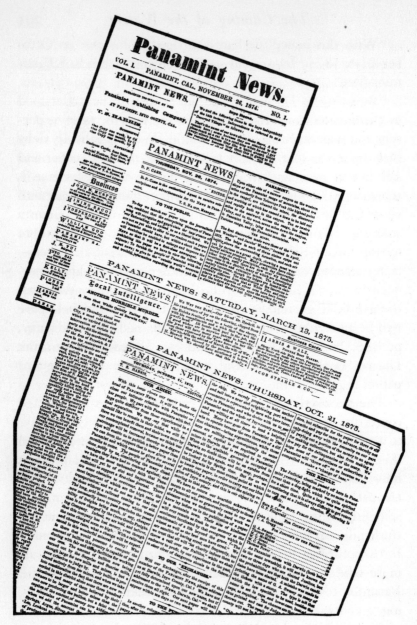

"Rushed no more editions."

"With this issue," he had composed at his case on October 21st, 1875, "closes our career under the head of Panamint *News . . .*

"By going to Darwin we do not wish it to be understood by the outside world that 'Panamint is a failure,' as some persons and papers are prone to conclude. We emphatically deny that this is the case. We merely propose to hoist anchor and sail into a near and neighboring port whose commerce is represented by men who have interest in Panamint, and with us desire to see Panamint and the whole of Inyo County succeed. Many have gone from Panamint to Darwin to operate because the ores there are of a different character— being argentiferous, while those here are milling—and consequently the same amount of capital is not required. When all the available ground is occupied in Darwin no doubt there will be an exodus from that place to this again, and this may, perhaps, be looked for by Springtime. But for us we think Darwin is the most central point at which to locate our paper. . . ."

The packmules bearing Editor Harris's press, types and concertina moved slowly down the canyon, taking five days to reach their new destination. The little building of the *News* was left behind to the rain, the waterspouts and the sun. Gales blew its windows in. Sage and rabbit brush grew up in the path to its door. The rust-gray phoebe, peering within, selected nesting spots in its rafters. The pack-rat built his trash-pile on the floor. From the canting chimney poured forth no more smoke, from its sagging portal rushed no more editions shouting sudden violence and swift acquittal. Panamint town could drink or shoot or sleep itself into oblivion.

For the *News* there was no springtime and no return.

JULY 24th, 1876, remained memorable for long in the intermountain basin. Its prelude came the day before, when a monstrous rainbag burst over the Diamond Range beyond Eureka, flung its contents down the ravines, swept upon and annihilated a camp of Italian woodchoppers, ripped through the hovels of fourteen Chinese who all were drowned or crushed to death but one, and moved on into Eureka to a wild accompaniment of firebells, steam whistles and falling walls. Warned by the oncoming racket, Eureka this time saved its collective lives if not its buildings.

Inyo county, California, discovered that it too had a part in the midsummer cloudburst carnival when a rancher named Lewis, in Bishop Creek Valley, heard a roaring in the Sierra canyons at his back and moments later beheld a wall of water several feet high and twenty to thirty feet wide advancing upon him like a brown-horse troop. To the ruin that was Lewis' there was added, shortly after, desolation and dismay between Owens Lake and Darwin. The Darwin tollkeeper ultimately found his wood supply, but it was in the lake six miles away.

All morning of that day, rain had been falling in the Panamints. Toward noon the sun came out for a moment; then the gray-black shadow-monsters began chasing each other over the hills once more. Fat black clouds, loaded with water from the Gulf of California, they prepared to split and dump their contents on the barrens beneath. But realization of an impending "waterspout" did not come to Panamint's remaining settlers until sudden bombardment in the high reaches was followed by an angry rumbling—the sound as of freight trains rushing through cosmos.

A miner, dropping his tools, came sprinting down from the

head of Surprise Valley, crying "Water! Water! To the hills! To the hills!"

Cabin doors flew open. What there was of the population of Panamint sprang out and made for the terraced trails as one man.

Then it came, charging down the main ravine that curved from the north. Rolling big stones and uptorn trees, the turbid crest met reinforcements from a southeast fork, rose high and surged on. From Telescope and Sentinel peaks newborn brooklets came. Plummeting from three directions down those two-thousand-foot walls, they joined.

The mill was on high ground. Its whistle began to bellow. To the crunch and snap of scraggly junipers, the tawny wave entered the village. It moved in a wide front upon the Surprise Company's masonry warehouse, was turned by that stout edifice, dashed angrily for frame structures just below.

Miss Donoghue's Wyoming Restaurant was protected somewhat by the neighbor masonry, but released a panic-stricken flight of benches and tables. Elias Cook's Pioneer Barber Shop sprang into air and flew apart. The cataract then enveloped the north footings of the dry-stone structures of Neagle's Block, battering in doors and geysering out through windows. It swirled into the store that had lately been Harris & Rhine's, washed through Dr. Wells' office and the long-deserted Bank of Panamint, toured Dave Neagle's forsaken splendors and caromed onward. Across Main Street it called for mail in Swasey's little post office, fell then with fury on Mrs. Zobelein's lingering little mart, and emerged with a rowdy pillage of hats, prints, parasols, overalls, linen and alpaca dusters, Broadway promenade shirts and moldering underwear.

With avalanching roar another sheet of water came racing down Sour Dough Canyon, stripping its sagebrush covering, ripping and raiding the shallow sepulchers. Still another, the swiftest and shrillest, flung itself down the ladder-like gulch that had been Maiden Lane. Its was a cleansing job and well it did it; never was an augean stable more thoroughly hosed out.

On down Main Street all the merging waters hurtled, bearing planks, trees, furniture; wrenching at cabin doors, sucking at foundations, tilting and demolishing. Munsinger's brewery, standing on a little island plot, tried vainly to part the waters; then its logs gave way. A quarter mile farther on, Wales' Bark Saloon went over with a crash.

Then, the brink at hand and the canyon walls pinching close, the river of the skies made its leap for the desert.

Four hours followed of incessant rain. When the sun finally released an ironic shaft, what was left of Panamint to its watchers on the hills was a lane of crazy timbers, cabins leaning groggily against each other, and stove-in walls.

Clean-scoured and radiant, the steeps loomed high above the wrack. Those chased-out inhabitants who still were sure the solid towers held millions returned to delve under their glittering façades again.

27

# THE COMING OF THE SHADOWS

Virginia City rebuilt swiftly. Had there not been fire and stock collapse before, followed always by sensational recovery? So the ensuing year found the public as convinced of the Bonanza's permanency as lately it had been of nearing exhaustion.

Then, in January of 1877, Con Virginia passed its dividend. And this time the bottom of the Bonanza was really in sight.

Once more they fought like entangled wildings—capitalists, hod carriers, clerks, scullions, screaming birds of paradise and their squawking mudhen sisters. Stocks that had sold up to $800 a share were on their way down to a stack of dimes.

May Day, 1877, was the blackest day in memory. "We are now visited by the most serious panic that ever swept over the stock market. The devastation and ruin wrought will be beyond all computation. No one can now say where this is to end," cried the *Alta*, which as much as any other agency had brought it all on. Huge blocks of all stocks came out and were thrown upon the market. Values fell off $20,000,000 in three days. "Never was there a more frightful state of affairs."

The seven brilliant years that started with Jones' discovery beneath Crown Point had come to their inevitable end.

Presently, "The shadow of hard times," requiemed Virginia City's *Chronicle*, "seems to be slowly creeping over the Comstock. Hundreds of miners, who have been out of employment for months, go about willing to perform any kind of labor for anything in the shape of money. A miner falls down a shaft and is dashed to pieces. A hundred rush to fill his place, content to have their faces fanned by the ill-wind that has blown him into eternity. They stand about the gambling tables watching the ebb and flow of other people's fortunes. A man with a dollar in his pocket is deliberating whether to buy flour for his family or put the money on the board to double up. He pushes it over on a card and sees it make its way into the dealer's till. Then he walks away with his teeth tightly clenched. You may see such sights nightly at any of the gambling houses. Strong, able-bodied men call daily at the dwellings of citizens, asking for food or the chance of a night's lodging."

The tidal wave of genuine ill fortune that fell upon the West this time knocked Jones flat with everybody else. It bowled him over, lifted him up and flung him down again.

But it was the malignant rush of those mysterious superheated waters, pouring in from some broken pipes in hell, that really overwhelmed him.

The water was seeping into Crown Point even as it was jetting into Con Virginia. Further exploration of the Crown Point hole became impossible. When its shares went down to a tenth of their former value, his fellow stockholders sought to oust the old officers. Between May 4th, when the rebels organized, and June 4th, when they took their case to law, Jones heeded the knell to all his present interests, and a messenger set off on the 600 miles to the Panamints.

There he delivered the order that stilled the little town's thirty stamps and ended its hopes forever.

It was the old story. Every pound of ore taken out of any vein decreases the life of the camp at its surface by just that much. A few years before, the town of Hamilton, off in the sagebrush to the eastward, had had nearly 8,000 population, Treasure Hill its 6,000, Shermantown 7,000, Swansea 3,000. Each was incorporated with mayor, councilmen, fire department; each had its daily newspaper. Hamilton now had but a few hundred; the fires of its furnaces were drawn; those who remained were waiting wanly for something to turn up. Treasure Hill had fifty people, Shermantown a single family, Swansea none. In one canyon of the Toiyabe range, near Austin, there were five dead towns without a single inhabitant.

John Percival Jones was definitely and grandly broke. The June morning that saw the start of the battle for Crown Point's reorganization saw also his railroad pass to the hands of the canny operators who had been steadily building their Southern Pacific southward and eastward and keeping out of Silver.

Shortly thereafter, Stanford-Huntington-Crocker engineers advanced upon the 2700-foot trestle that, in anticipation of Panamint's ores without limit, had pushed so bravely out into Santa Monica Bay. As heavy sledges rang, and the pier came down plank by plank, the aspiring little city behind it shrank visibly. The Jones touch, Midas-like, had made a town out of a rocky gorge up in the Panamints and this town by the sea. Now that Midas had lost his touch, Panamint was reverting to the gorge and Santa Monica to its sheep pastures and sandblown bluffs. As from Surprise Valley, so now from these palisades above the tide the population turned and fled. In time it returned, to Santa Monica at least, to make on those lovely eminences a region of life, gaiety and urbanity; but Santa Monica's dreams of commercial supremacy crashed with the fall of Panamint. For years thereafter the stumps of Jones' long pier stood in the surf, dolorous reminders of the bonanza that never had been, and the boom that had died.

Colonel James V. Crawford, disappointed by the fall of his L.A. & I., was soon mollified by a brilliant post as Superintendent of Railways for Japan. To his abandoned Cajón project back of San Bernardino the wild toyon berry and the mountain lilac swiftly returned. The narrow gauge grade reverted to a mule track and the elaborate tunnel caved in. Yet other decades brought new activity, and though Panamint never got its connection, Cajón canyon did get its rails; Santa Fé and Union Pacific whistle through Engineer Crawford's hotly won pass today.

The promised telegraph likewise never reached Panamint. The first message ever to be carried across Inyo county by bridled lightning arrived at Independence from Caliente on January 23rd, 1876. It was borne over the wire of the Inyo Consolidated Calithumpian Telegraph Company, and the message read: "I'm cold and dry—send me a cocktail."

R. C. Jacobs put some of his Panamint gleanings into a city square in Oakland. He built a big house at the corner of Ninth and Union Streets, called the surroundings "Panamint Block," and invited all old associates to come and help colonize it. For himself, the rainbow trail still called and he was soon off for Mexico.

Up in Inyo's spaces, the firm of Small & McDonald came to a parting, the liquidation being speeded by McDonald's gun. Loser by a matter of a split second was John Small.

Jack Lloyd, who had tooled so many Panaminters back and forth across the sagebrush, was himself a passenger in Billy Balch's stage from Mojave for Darwin when a robber poked a shotgun out of a creosote bush on February 14th, 1877, and sent without warning a heavy leaden valentine. Lloyd had been asleep in the hind boot. He stuck his head up just in time to receive the whole charge. Billy drove on into Darwin with what was mortal of the jesting reinsman. Lloyd had evidently been mistaken for Wells Fargo's shotgun messenger Bob Paul, later a famous Arizona express guard and sheriff, who had missed that ride.

Señor Chávez rode away from Inyo trails to early extinction. He was seen at the Lievre rancho near Los Angeles in the summer of '75, where he halted and asked for supplies. As a gun in each hand backed the señor's visit, he was courteously provisioned. In the autumn he appeared at a rancho on the Gila sixty miles above Yuma, where he found employment under an assumed name as a horsebreaker.

Heavy rewards hung over him, and they inspired a suspicious herder to report his identity to two gringos. These advanced on Baker's Camp resolved to bring Chávez back with them alive or dead. They made a quick last-minute choice in favor of a Chávez dead, gave him two barrels in the back and then ordered up his hands. Chávez died without speaking. One of his extinguishers then took the head, preserved in spirits, to San Francisco and presented it to the authorities. Though admitting that the Vásquez lieutenant looked better in a jar than in the saddle, officialdom pointed out that the term of the rewards had expired while the evidence was en route. For exhibition purposes it was discovered that by shaking the jar, causing ripples and refractions, the gran ladrón could be made to smile or scowl at will.

Pat Reddy passed through all the vicissitudes at both Darwin and Bodie, and in subsequent years could be found defending the guilty with eloquence anywhere from Carson City to the Coeur d'Alenes. Blunt, brave, unscrupulous, he marched through the changing scene accompanied by a bodyguard to offset the handicap of his missing arm. An appreciation of the man was voiced in 1883 by the Bodie *Free Press:*

"Hon. Pat Reddy, Joint Senator from the Fourth District, seems to have his hands full. Is floating mining schemes all the way from Bishop Creek to London; defending all the criminals from Inyo County, Cal., to Carson, Nev.; has a contract on hand to lick the warden of San Quentin prison or any Republican member of the legislature who may tread on the tail of his coat; keeps a law office in San Francisco, also one each in Carson and Bodie, and holds himself in readiness to fight, buck or gouge a nest of wildcats, or overturn the federal administration, just as things come handy to him." Ultimately the busy man found time to run unsuccessfully for the California governorship.

E. P. Raines, as he rode down Surprise Canyon at the height of the boom, cast about for something new on which to exercise his wits. He had in his wallet, besides J. P. Jones' check, the clippings concerning that 1868 stagecoach robbery.

Thus it was that, in due course, an Oakland man named Bailey was approached by a smooth-speaking acquaintance who had a curious tale to tell. Following that large-scale Idaho holdup of years before, several men had been arrested and one had been locked away for a long term on another charge in the Nevada State Prison. Finding himself ill, this desperado confided to a turnkey that he had buried a large part of the stolen $64,000, and gave the kindly official a map. Then died. The turnkey lost no time resigning his post and hurrying to the spot marked X, where, sure enough, he dug up many bullion bars. These he took to Reno and buried again.

Then the ex-turnkey, according to Bailey's informant, found himself in a predicament. He had lost his job and he didn't dare cash this Wells Fargo bullion. Bailey was given a chance to buy these bars at a bargain price and was introduced to the rest of the conspirators—one of whom was E. P. Raines. Bailey found himself much awed by their talk. Here, he was led to assume, were members of the very band that once had touched up Baldy Green on the Geiger grade, stealing his coach and six; that had pursued Bill Blackmore's leaping vehicle; that had carried off a whole wagonload of bullion coming from a mine near Austin; that had conducted wholesale mine and mill robberies, melted and re-stamped the metal, and shipped it to distant agents for sale at the mints. Bailey felt quite flattered to be an associate of such large-talking gentlemen.

Here the story, as told later to a San Francisco jury, grew complicated. Bailey, having put up his money, demanded to be shown one of those Idaho bullion bars. He was promised this privilege, and a rendezvous set on an Oakland ferryboat. And sure enough, there he was shown a brightly yellow, exceedingly heavy bar. But when Bailey reached for the bar, by unlucky accident it dropped overboard.

All this set Bailey to thinking, and he concluded that he was being jobbed. He went to the police about it. On October 26th, 1876, Raines was arrested and charged with the manufacture of the gold brick shown to Bailey on the ferryboat. Eleven false bars were found in his possession and he was accused of running a neat business in this form of counterfeiting.

It took fast talking on Raines' part to get out of that fix. But the brick that Bailey had paid for was at the bottom of the bay, nobody could prove that it at least wasn't solid gold, Raines was as debonair as ever, and the upshot was that five months after the arrest, "Mr. E. P. Raines," the Prescott *Arizona Enterprise* was reporting, "was one of the arrivals from San Francisco by Thursday's stage. Mr. Raines has been interested in some of the largest mining operations on the Coast." A year later: "E. P. Raines is developing a series of mines in this locality. At eight feet he has met with a very fine vein of ore. . . ." A vein, one suspects, which did not suffer any shrinkage when Raines described it.

Editor T. S. Harris at Darwin set up his press and types with some success, his *Coso Mining News* becoming a well-printed sheet seldom lacking for material about knifework, gunfire, and sudden demise. When Darwin faded, Harris again packed his press and types and set sail for rising Bodie whither Bill Stewart, Pat Reddy and other old acquaintances had preceded; the camp concerning which undertaker Ward was insisting "There never have been twelve bodies un-buried in Bodie since I have been here. The most at any one time was seven"; and of which a contemporary journal was reporting: "The growth of the town has no parallel in the history of mining. Fine residences, saloons, business houses, brothels and cabins are in a motley jumble as to location. There are 47 saloons and 10 faro tables. This is not a disparagement of the district, but an evidence of its prosperity."

But the violence that had been life's pattern at Panamint and Darwin continued to influence Harris' destiny. The *Standard* dimming with the eclipse of Bodie, he drifted to Los Angeles and helped found the *Evening Republican;*

became involved in a row with his managing editor, fell back on the Panamint-Darwin-Bodie formula and let fly with the contents of a six-gun. For this reversion to wilderness ways Harris was sentenced to a year in San Quentin prison. Upon release he tried to rehabilitate himself, set up a little printshop in San Francisco on Ellis Street, found the lady-card definitely turned down, and in 1893 put a .44 pistol bullet in his brain, leaving a note that carried the salutation to the world, "I have had a great time."

When Superintendent Messec received instructions to wind up the Surprise Valley Company's affairs, he found himself possessed of about one hundred burros. These he turned over to one of the camp's Indian laborers, the worthy known as Jake, allotting him one-third of the animals for his trouble as caretaker.

A year or two passed, the company's share of jacks was disposed of to another party and the new owner came and rounded them up, leaving Jake with thirty-nine beasts for his trouble. Indian Jake, his clear title established, promptly began trading them to his kinsmen for a harem—"one jack, one wife; bery good." Thoms, the purchaser of Jones & Stewart's seventy-eight animals, "brought all of his share of of the jacks away," reported the *Independent*, "not being in-clined to speculate, like Jake."

Just before his leave-taking, Captain Messec played host once more to three former Panaminters who came up the canyon. Their names were Brannan, Murphy and William Raymond. They were on the march from Darwin for Resting Springs, which lay across Death Valley. Messec advised the trio to drop down to Bennett's Well by way of the granite wall back of Panamint, and thence to strike by night for Furnace Creek. This was a beaten track eastward. The travelers chose instead to short-cut directly southeast. It was forty-five miles to the next water by that route, across an August desert. They left Bennett's Well in the evening, pushed on all night, found themselves out of water at ten o'clock the next morning and realized they had ignored some good advice. They concluded to return.

Midday found them in the valley's hottest part. At three o'clock Raymond halted exhausted. Murphy gave out soon after and sought protection by burying himself in the sand.

Brannan struggled on. He made the fifteen miles back to the base of Telescope Peak. The borax men had long since left their camp in that vicinity, driven as from an inferno. The stout-hearted Irishman climbed the slopes and came to an Indian wickiup. There he obtained native help and two horses, and moved back over his late track. Murphy was found. He had maintained himself alive in his self-made grave for a day and a night without water, in a temperature that was beyond conception.

The rescuers then shouted for Raymond. He was nowhere to be seen. He had wandered in a maniacal daze for the horizon, or succumbed in some shallow gully on the salt-white plain.

The Comstock-born panic of 1877, if it flattened J. P. Jones, Panamint, Santa Monica and the L.A. & I. along with all the rest, was after all but one more gale to an oak that was used to hurricanes. When the great fire laid waste Virginia City, "J. P." was in the forefront of relief activities, raising funds in San Francisco for the stricken miners and urging an assessment on every Comstock share for the shelter and provisioning of the destitute. Within a year of the wind that carried away so much more than his hat, Jones had recovered a full quarter of his losses.

Then in 1878 the blasts leveled him again. For at least the third time in his life he picked himself up with good humor. Meanwhile two of his assets, or perhaps they were the halves of one, remained unshakable. Nothing could damage the popularity or the unvarying political success of this even-tempered man. During the course of his service he broadened into a real Senator of the United States, instead of a mere Senator for Nevada. He served for thirty consecutive years, issuing tremendous Monetary Reports which everybody admired if few read, and divided his time between the capital, the rowdy silverlands, exhilarating San Francisco and his beloved if disillusioned Santa Monica.

A decade after Tom Fitch auctioned off the sea view and the sunset around Shoo Fly Landing, a new boom developed in those adjuncts to delightful living and Jones as owner of the land beneath found himself again in opulence. This time the Senator built a home vast with wings and bays on the bluff above Santa Monica's beach, named it Miramar, and for many years it was a center of southwestern gaiety and hospitality.

Plunger and promoter to the end, Jones among innumerable other undertakings helped to organize and develop the great Treadwell mine in Alaska. When nearing eighty he retired from public life, withdrew to expansive Miramar, and in 1912 at Los Angeles passed peacefully from this always lively, perpetually variable and generally agreeable plane.

He had lived to hear the whistle of a railroad locomotive down across the wastes from Salt Lake to Los Angeles. But he died too soon to behold that other vindication of his dreams —Boulder Dam. Yet this mighty work might well wear a bronze tablet on its concrete face to the man who, sixty years too soon, sent engineers and surveyors into the field to trace canals and man-made lakes that should water the deserts of the Colorado River and make them bloom.

The great $5,000,000 recovery suit of English stockholders over the Little Emma fiasco ended in victory for Stewart and Trenor Park—their intentions were found honorable if their judgment insecure. The mine, after being declared worthless, then confounded all—stockholders, managers, and promoters who had sold it out—by discovering in its depths some $15,000,000 in authentic silver. Capricious are the ways of the pale goddess. Park, who had wept when he was vindicated on charges of being a swindler, wept anew when he learned what he had swindled himself out of. He died at sea before the full bulk of it was revealed.

Bill Stewart, broke and cheerful about it, was rebuilding his law practice at Virginia City when the stamp mill at Panamint went silent. For the next nine years his big frame and leonine beard moved through the mining litigation of the West.

But the speculative flame burned on. As Panamint slid into the shadows, or washed away down its canyon, and Bodie was getting ready for its tumultuous revival at the north, Bill Stewart again was shouldering his way through throngs of roughs, loafers, sagebrushers, bullwhackers, sourdoughs, promoters, and plain miners, tossing his accumulations into the holes on Silver Hill, and declaiming for the San Francisco press with inveterate optimism: "The Comstock may as well gracefully release its hold on its laurels and resign its claim to leadership in favor of Bodie."

When Bodie's nights grew still and Tombstone's raucous, Stewart moved on to the latter. In the wild cowboy capital on the Arizona border he found many an old crony. Dave

Neagle was there, wearing the badge of a chief of police, and Allen and Tough Nut streets were graced with "many of the 'gentlemen' who had lived in security at Panamint in defiance of the law before they secured full freedom by dividing the purchase money of their claims with Wells Fargo"—the words are Bill Stewart's—and who had "retired to Arizona after mining operations ceased at Panamint."

Stewart and Neagle again touched shoulders in history when the latter, chased out of Butte, Montana, after a shooting, got a job as bodyguard to United States Supreme Court Justice Stephen J. Field. The Justice had figured in a knotty case, the effort of a young woman to force William Sharon to live up to a marriage contract. The sensational suit by "Sharon's Rose" had brought face to face many characters of the rough old West, including Bill Stewart, Sharon's former political rival, sitting beside him as one of several counsel; and David S. Terry in similar capacity at the elbow of the languishing plaintiff. Though Stewart's old legal adversary lost his case he won his fair client, married her, and in a subsequent phase of the dispute was clapped into jail by Justice Field for contempt.

Terry vowed to have the jurist's life. On August 14th, 1889, the Justice was seated in the railroad dining room at Lathrop. Terry entered. He saw Field, walked up between the tables, and made a threatening move. Whereupon Dave Neagle of Pioche, Panamint, Bodie, Tombstone and Butte swiftly drew the gun that had so often compelled peace in tougher places than Lathrop, and the roar of it ended Terry's career forever.

Litigation and Tombstone were synonymous, and while both lasted, shape and fullness were restored to Stewart's purse. Meanwhile Jim Fair and his Bonanza profits had replaced Sharon in the Senate. In '86, after two terms out, Stewart astonished Fair and his hundred million dollars by snatching back the seat he had once relinquished to Sharon, and Sharon had surrendered to Fair. For the next eighteen years Stewart again served beside Jones, while the beards of both increased in length and snowiness. His last election, that of 1898, was seized in the approved Nevada manner. His opponent appeared to have the necessary pledges when the legislature convened. Then it was discovered that one less vote in the Assembly would throw the election Stewart's way. Just before the ballot was taken, a fast vehicle left Carson City forcibly bearing a young legislator out of the state and into a handsome corporation job at San Francisco. His abrupt departure up the grade left the old warrior supreme.

But the years were slipping fast, and Stewart's fortunes with them. At the turn of the century he was busted again.

One resource remained: Stewart Castle. In recent years it had been rented out to the Chinese legation. Stewart had not approved of his tenants' style of housekeeping and had ejected the embassy in wrath. Now he decided to let the internal finery go.

For three days the castle contents were auctioned off to gaping Washington. Forth and out went the big mirrors, the oil paintings that had been bought by the yard and the tremendous walnut bedsteads. Forth went the table silver pounded out of Comstock and Panamint metal. Down the imposing staircase and out beneath the gas-lit arched entrance went the outmoded bric-à-brac. Stewart and a new young wife lived for a time on what these relics brought, while the famous castle stood forlorn with its stucco falling and its turrets looking down on Dupont Circle with a seedy leer.

Finally William A. Clark of Montana, who had succeeded both Stewart and Jones as the mining Croesus in the Senate and Jones as actual builder of that railroad from Salt Lake past Death Valley to Los Angeles, bought the Stewart landmark and pulled it down to improve his view.

Stewart stepped from the national forum for good and all on March 3rd, 1905. His wealth was gone, but the sagelands were abloom with spring and there was spring in his hale old breast. In May the veteran gladiator hitched four horses to a camping wagon, loaded in his law books and set off across the deserts. In the Bullfrog District just off the northeast corner of Death Valley, 3,500 feet above the sea, he set up his tent and once more went to work. What does it amount to to be out of fortune when you're eighty years young and you've found a boom camp again that's going to be "bigger than the Comstock"?

Important mining suits were won anew by the rawhided old fighter and with them came big fees. Speculation added to his success. For the dozenth time he had caught success by the skirt-hem and there was an hour or two when his fortune, on paper at least, was accounted one of the largest in Nevada. So he built himself the biggest house in Bullfrog. Through its front door he could see the Amargosa Range, and through a notch in that crestline the far blue tip of Telescope Peak above its companion Panamints. Sight of that familiar upthrust kindled old memories, and while residing here Stewart dictated his reminiscences.

But the Panamint Mountains and the camp to which they had given name were covered now with haze. To the dinful scene of thirty years before he tossed off but a dozen pages. When suddenly, just at the end of these literary efforts, his assets once more turned to vapor. The Bullfrog District's pet stock, Montgomery-Shoshone, slid from $20 a share to little or nothing and William Morris Stewart was smashed again. And this time he turned his back on the desert forever. The house that was the pride of Bullfrog slumped under sunheat and cloudbursts, its roof caved, and it joined the ranks of the eternally tenantless.

Washington then knew Stewart for about a year more, a gallant patriarchal figure. His tall frame, long white beard, ruddy cheeks and rakish soft-brimmed hat made him a sort of western incarnation of Santa Claus, and children missed him when he no longer passed their play. He died suddenly at a Georgetown hospital in 1909, leaving a law library valued at $25, a $200 horse and wagon, and $25,000 in debts.

When the Surprise Valley Company passed on, a few miners who had no place else to go remained at Panamint with the memories of days and nights that had flared and dimmed. One of these "chloriders," an old county register shows, was James Bruce. Salvaged from his wounds, he returned and occupied his later years as a simple miner, though the days of high play were gone and Main Street was no longer any sort of a shooting gallery.

Far up Sour Dough Canyon, higher than all the rest of the fifty-seven, Bob McKenny's tiny plot clings to the rocky hillside. Though the other graves have long returned to nature, the wooden palings about McKenny's resting place have been occasionally renailed and have stood through the years intact. Its weather-worn headboard reads "In Memory of Robert McKenny. Died April 17, 1875, aged 30 years 3 months 13 days." One wonders: did Jim Bruce's hand preserve the sepulcher of the man who so nearly sent him to Sour Dough in his place? Only the bending junipers know.

Finally Jim Bruce himself heard the call of new places, and the turn of the century found him running a roadside tavern on the outskirts of Goldfield.

To the old-timers who made their homes in Panamint's sagging hovels there remained the billiard table once brought up the canyon by Mr. Neagle. For thirty years it stood in what was left of the Oriental, a raceway for mice, while its ivory balls grew increasingly elliptical and the cues went spiral in the rack. At last, with the rise of a new camp at Post Office Springs down on the edge of Panamint Valley, the vasty table was remembered and once more borne the length of Surprise Canyon—this time downward, to bask in the lights of Chris Wicht's resort at Ballarat.

In the course of the wheeling years Panamint has had a few sporadic revivals. Silver is still there, though tightly held in the cliffs that once drew two thousand men across the deserts. After some years the "big mill" burned down and a ten-stamp affair was constructed of the remains. Eventually this too grew still.

Down in the mesquite thicket on the edge of Panamint Valley where Chávez once camped, Indian George, the guide who led prospector S. G. George nearly to Surprise Valley in 1860, is still living. Chieftain of the nineteen aboriginals who survive in a country 150 miles long by 100 miles wide, he claims an age approaching the century mark, recalls clearly the turbulent days and nights when Panamint was a lively camp, keeps Jack Lloyd or Billy Balch's abandoned old stagewagons for chicken-roosts, and will tell you how he watched the original Forty-nine party make their doleful crossing of Death Valley when the West was young. They must have been frightening creatures with their monstrous oxen, creaking wheels and awful face-whiskers. The little Indian and his daddy could have led them to spring water and safety. "But for what—to get shot?"

High above, Telescope Peak invites the cloudbursts. At its feet, Death Valley and Panamint Valley burn in the sun. Visitors like ourselves occasionally climb what is left of Bart McGee's steep wagonway to poke among the cabins of the deserted town. Appropriators have made off with the doors and windows and most of the iron machinery and champagne bottles. Panamint sleeps on.

Jim Bruce will not be back today to this cabin of cold hearth and shattered roof. Martha Camp will not doff her whalebone corsets. Sophie Glennon will not hold her lamp tonight.

We draw shut the door.

# INDEX

first funeral, 96, second, 167; Barstow buried, 177; McKenny buried, 267, 313.
Southern Pacific Railroad, 65, 212, 214, 216–17; 270, 277, 286.
Spear, Colin A., shot, 206.
Spratt, Jack, shoots highwayman, 253.
Stanford, Leland, railroad builder, 67, 212, 214–15, 219, 274, 299.
Stetefeldt, C. A., mining report, 84.
Stewart, Robert L., 285; prospector, 23–8, 37, 44; departs, 97.
Stewart, William Morris, U.S. Senator and mine owner, 11, 19, 33, 50, 65, 72, 83, 115, 117, 119, 124, 144, 145–6, 151–3, 156–7, 160, 180, 204, 206–7, 209–10, 216, 230, 234, 261–7, 278–9, 281–2, 285, 308–11; sketch, 102–14; buys robbers' claim, 200; escapes Chávez, 267; quits desert, 312; death, 312.
Stewart Castle, description (1874), 112–13; finery sold, 310.
Stock & Exchange Board, 38.
Sullivan, Georgina, 145, 209.
Sullivan, Jerry, shoots at Robertson, 169; fells Govan, 179.
Surprise Canyon, 34, 45, 48, 63, 76, 82, 84, 87, 89, 96, 156, 161, 163, 180, 207, 214, 219, 253, 256, 263, 288; description, 15, 16; relocated, 26.
Surprise Valley, 16, 30, 34, 47, 82, 93–4, 102, 123, 145, 157–8, 191, 222, 227, 286, 288, 313.
Sutro, Adolph, 62.

Tait, Sam, resort keeper, 89, 90.
Telescope Mining District, 16.
Telescope Peak, 23, 31, 42–3, 69, 78, 89, 188, 220, 224, 227, 287, 295, 306, 314; view from, 13–4; named, 14.
Tempiute Bill, Indian, 98; lynched, 99.
Temple, F. P. F., banker, 46, 48, 65–6, 145, 207, 284–5.

Temple & Workman, 46, 48, 66, 285.
Terry, David S., 107, 309.
Tevis, Lloyd, 154.
Texas & Pacific Railroad, 65, 212.
Tipton, Daniel G., locator, 32, 130, 153.
Towne (Townshend), Forty-niner, 6, 10, 12.
Twain, Mark, 11, 111.

Vanderlief, ———, promoter, 41, 45, 46.
Vásquez, Tiburcio, bandit, 245–6, 248, 258.
Virginia City, 39, 60, 83, 89, 94, 106, 115, 133, 137, 143, 148, 262, 308; Bonanza revealed, 155, 173, 204, 207, 209; earthquake, 210–11; fire, 286–7, 306; rebuilds, 297; *Chronicle* of, 144, 298; *Territorial Enterprise* of, 231.
Visalia & Inyo Road Co., 120.

Washington & Creole mine, battle, 71.
Welch, E. P., killer, 234; escapes, 238, 240.
Wells, Fargo & Co., Express, 70–1, 79, 80, 129, 156, 164, 193, 195, 198–200, 278, 301, 303; refuses to deal with Panamint, 157.
Wheeler, Lieutenant, explorer-geologist, 95.
Whitney, Mount, 14, 83.
White Pine, 71, 74, 89, 92, 192.
Wicht, Chris, xi, 313.
Wilson, John (John Curran), locator, 33; seizes Defiance mine, 229, 231–2.
Wilson, William, prospector, dies in desert, 95.
Wolsesberger, Uncle Billy, 92, 222.
Wyoming mine, 160, 223, 278.

Yager, Fred, resort keeper, 2, 179–80, 289; mirror arrives, 181.
Yellow Jacket mine fire, 53–4.

Zobelein, Mrs., storekeeper, 92.